CHINESE ECONOMIC POLICY

ECONOMIC REFORM
AT MIDSTREAM

Professors World Peace Academy Books

World Social Systems Series
General Editor
Morton A. Kaplan

OTHER BOOKS AVAILABLE

THE SOVIET UNION AND THE CHALLENGE OF THE FUTURE

Editors
Alexander Shtromas and Morton A. Kaplan

Vol. 1 *Stasis and Change*
Vol. 2 *Economy and Society*
Vol. 3 *Ideology, Culture and Nationality*
Vol. 4 *Russia and the World*

CHINA IN A NEW ERA

Series Editor
Ilpyong J. Kim

Chinese Defense and Foreign Policy
Editor June Teufel Dreyer

Chinese Economic Policy
Editor Bruce Reynolds

Chinese Politics From Mao to Deng
Editor Victor C. Falkenheim

Chinese Science and Technology
Editor Denis Simon

also planned:
LIBERAL DEMOCRATIC SOCIETIES

CHINESE ECONOMIC POLICY
ECONOMIC REFORM AT MIDSTREAM

Edited by
Bruce L. Reynolds

Series Editor
Ilpyong J. Kim

A PWPA Book

PARAGON HOUSE
New York

Published in the United States by
Professors World Peace Academy
481 8th Avenue
New York, New York 10001
Distributed by Paragon House Publishers
90 Fifth Avenue
New York, New York 10011
Copyright © 1988 by Professors World Peace Academy
All rights reserved. Except for use in reviews, no part of this book may be reproduced, stored in a retrieval system, or transmitted in any form or by any means, electronic, mechanical or otherwise, without the prior written consent of the publisher.

A Professors World Peace Academy Book

The Professors World Peace Academy (PWPA) is an international association of professors, scholars and academics from diverse backgrounds, devoted to issues concerning world peace. PWPA sustains a program of conferences and publications on topics in peace studies, area and cultural studies, national and international development, education, economics and international relations.

Library of Congress Catalog-in-Publication Data

Chinese economic policy /
edited by Ilpyong J. Kim and Bruce L. Reynolds
  p. cm.
  "A PWPA Book."
  Includes index.
  ISBN 0-943852-69-2 (v. 1): $29.95.
  ISBN 0-943852-70-6 (pbk. : v. 1): $14.95
  1. China—Economic policy—1976-
  I. Kim. Ilpyong J., 1931- .
 II. Reynolds, Bruce Lloyd, 1944- .
HC427.92.C451458   1989
338.951—dc20
89-9423 CIP

Table of Contents

Preface	*Ilpyong J. Kim* *University of Connecticut*	ix
Foreword	*Bruce L. Reynolds* *Union College*	xi
Introduction	*Ezra F. Vogel* *Harvard University*	1
Chapter 1	Dilemmas of Partial Reform: State and Collective Firms versus the Rural Private Sector *David Zweig* *Tufts University*	13
Chapter 2	Still on the Collective Road: Limited Reform in a North China Village *Thomas B. Gold* *University of California–Berkeley*	41
Chapter 3	The Chinese Village Incorporated *Jean C. Oi* *Harvard University*	67
Chapter 4	The Chinese Economy in the New Era: Continuity and Change *Robert F. Dernberger* *University of Michigan*	89
Chapter 5	Property Rights, Economic Organization and Economic Modernization During the Economic Reforms *Ramon H. Myers* *Stanford University*	137

Chapter 6	Rural Factor Markets in China After the Household Responsibility System Reform	169
	Justin Yifu Lin	
	Development Institute, Beijing	
Chapter 7	The Impact of Reform on the Saving and Investment Mechanism	205
	Bruce L. Reynolds	
	Union College	
Chapter 8	City, Province and Region: The Case of Wuhan	233
	Dorothy J. Solinger	
	University of California–Irvine	
Chapter 9	The Political Economy of Regional Reform: An Overview	285
	Victor C. Falkenheim	
	University of Toromto	
The Editors		311
Contributors		312
Index		315

LISTING OF TABLES

Development of Small Business in China, 1981–1986	18
Development of Private Business in Nanjing, 1978–1985	18
Individual Business Employment Structure, Jiangpu County, 1978–1986	19
Private Sales as a Percentage of State Agricultural Trade, Nanjing, 1983 and 1985	32
Fengjia Village's Marketing System	47
1985 Income for Fengjia Village	48
1985–1986 Grain Production	48
Per Capita Income	49
Average Annual Rates of Growth of Net Material Product	93
Structure of Production	97
Structure of Employment	98
Rates of Accumulation or Investment	100
Foreign Trade Participation Ratios	101
Employment of Labor Force by Type of Ownership	103
Retail Sales by Type and Ownership	104
Sources of Growth: State Owned Industry	110
Proportion of Material Consumption to Gross Output Value by Sector	111
Principle Financial Items of State Owned Enterprises	112
Per Capita Consumption of Consumer Goods	114
Per Capita Housing Space and Savings	114
Per Capita Levels of Consumption	115
Ratio of Peasants to Non-Agricultural Residents Per Capita Consumption	116
Distribution of Income	118
Land Endowment in Each Province	176
Land Endowment in Each Prefecture, Anhui Province	177
Scope of Land Transactions in Some Areas	179
Labor-Hiring in Wu County and Yangshi County	185
A Survey of Cash Expenditure and Types of Credit	193
Agriculture's Role in the Economy	214
Intersectoral Sales	215
Distribution of Gross Domestic Savings	216
Sectoral Allocation of Capital Construction Funds by State-Owned Enterprise	222
Investment in Agriculture, 1985	224

PREFACE

This volume, composed of nine provocative chapters by prominent China specialists, analyzes the extent of economic reform in China to date, particularly in rural areas, and the impact of reform on economic activity. Initially, these chapters were presented at the Third Congress of the Professors World Peace Academy (PWPA) at Manila, the Philippines on August 24-29, 1987. Under the conference theme of "China in a New Era: Continuity and Change," the congress addressed issues concerning the past, present and future development in China.

A total of 12 panels were organized with the participation of more than 98 China specialists from the five continents. Their research papers addressed various aspects of Chinese politics, economic development, law and society, industrial and rural reforms, and foreign and defense policies. The papers were subsequently revised by the authors and collected into five separate volumes, of which the present volume, *Chinese Economic Policy: Economic Reform at Midstream*, is the second book to come off the press.

Volumes on politics, economics, comparative reforms, science and technology, as well as Taiwan, will follow in sequence. It is our hope that these volumes will contribute to our understanding of China as it makes the great leap toward the 21st century under the "four modernizations" program. The organizing committee, which I chaired, attempted to organize the panels to be both interdisciplinary in scope and international in makeup.

I would like to express my heartfelt gratitude for the financial support the International Cultural Foundation provided for the convening of the Manila conference and the publication of these volumes. Gordon Anderson, General Secretary of PWPA, and his staff members as well as Kevin Delgobbo, the program coordinator, and Robert Brooks, Publications Manager, have worked tirelessly for the conference and the publication of these volumes. However, the views expressed in these volumes are the sole responsibilities of the chapter writers, and do not represent the views of the organizers, sponsors, or editors.

                                Ilpyong J. Kim, Series Editor
                                Storrs Connecticut

FOREWORD

Bruce L. Reynolds

As the 1980s draw to a close, the economic reforms which began in China in 1978 have clearly had a powerful impact. A decade of rural reform, by giving individual farmers the residual fruits of extra effort, more than doubled the growth rate in agriculture. Agricultural output per worker rose 250 percent, in part because 80 million underemployed peasants left the land for non-agricultural employment. This massive transfer of labor led, in its turn, to explosive growth of rural industry. Foreign trade also mushroomed in these years, and per capita incomes rose rapidly. In 1989, China could look back on ten years in which its GNP growth record, more than 10 percent per year, far outstripped that of any other country in the world.

But as China enters the last decade of the twentieth century, economic reform is incomplete. Agricultural product prices are still heavily controlled by the state, and factor markets for land, labor and capital in rural areas are in their infancy. In industry, the state-owned sector still dominates, and within that sector, the planning bureaucracy dominates market forces. In foreign trade, free convertibility of the currency is at best a distant goal. Most important, there is not yet a clear commitment to the use of market-determined prices as the fundamental guide throughout the economy, and with the sharp inflation of recent years, such a move appears less and less feasible.

This volume examines, in detail, the characteristics of this hybrid economic system: Chinese economic reform at midstream. Most of these papers (in particular those by Zweig, Gold, Oi, Lin and Solinger) are detailed case studies of particular sectors or regions. Some (Dernberger, Falkenheim, Myers and Reynolds) are broader in scope. But each paper asks, in one fashion or another, the same question: what has been changed by reform? What remains the same? And which element dominates: change, or continuity?

Of the non-Western countries which have moved through a growth acceleration and emerged as middle- or high-income industrialized nations, almost all lie in East Asia. This fact has led many scholars to focus on an East Asian development path or model. Inevitably, China's growth performance in the 1980s has produced speculation that China, too, will now follow that path. Ezra Vogel's introduction to this volume poses that question: will China become another NIC (newly-industrialized country)? The paper, originally prepared as a welcoming speech to the first plenary session of the Manila conference, lists nine characteristics or special circumstances which the East Asian NICs enjoy. Vogel then points out that China lacks several of those advantages. Still, he predicts that certain regions, in particular South China, may be "someting like" the NICs by the early 21st century.

The first three chapters in this volume, by David Zweig, Thomas Gold and Jean Oi, present case study data on the impact of reform on rural China. The research sites differ markedly, both geographically and in the extent to which economic reform has affected the rural economy. Zweig describes a relatively urbanized county near the city of Nanjing, in the Yangzi River region which has experienced rapid rural growth and relaxed government controls. Gold's village, by contrast, in Shandong Province, seems nearly untouched by the rapid decollectivization which has occurred elsewhere. Jean Oi's interviews cover fourteen sites, not create problems for its economy, that it could absorb the products from these countries without unfortunate impact on its own economy. It was a special historical period when powerful countries in the world had a special relationship to developing countries.

A second characteristic was the basic strategy of these countries: the positive use of changing comparative advantage. The old Ricardian notion that has dominated Western economic thinking about comparative advantage was based on the example of England with a fairly cool climate and rolling hills good for sheep herding. It was appropriate, Ricardo felt, that the English herd sheep and make wool; Portugal with a warmer climate, could grow grapes and make wine. It would benefit everyone if the wine made in Portugal could be sent to England and the English wool sent to Portugal. That each country would respond to its current advantage was the classical Western notion. Japan pioneered in rethinking that notion. Japanese leaders believed that because Japan had so few resources, it could not achieve prosperity based on its comparative advantage. It would use one level of comparative advantage as a foundation and preparation for the next. First the Japanese used cheap labor to manufacture textiles and other cheap goods. During that time they were preparing for a later stage when they would manufacture light electronic goods.

Capital earned in one stage would be used for heavy capital investment in the next stage. The Japanese moved from ship building to steel, then to automobiles, and then into high technology. Now they are moving into research information. The concept of changing comparative advantage, rather than static comparative advantage, was central to those countries.

A third point is the role of the government to guide that process. All three relied on elite leadership, on meritocratic bureaucrats who achieved their position on the basis of examination and proven ability. This situation was consistent with the East Asian classical tradition, yet it had a modern focus because elite bureaucrats with modern training guided the process. Industrial and commercial work was done by independent private business. But there was a close relationship between businessmen and bureaucrats. Bureaucrats did not run the economy; they merely guided it.

A fourth characteristic was that these governments have proved to be strong, stable, and until recently, fairly authoritarian. As much as we might like to idealize the successes of these countries, we must look at the situation objectively. The top leaders of these countries did not lead in Western style democratic ways. Even in Japan, post-war leaders like Yoshida Shigeru, Kishi Nobusuke and many other LDP leaders did not follow the Western version of open democracy. Rather the strong authoritarian leadership in these countries served to control people who were inconvenienced and who had to make sacrifices during the process of development.

A fifth characteristic is that these countries were strongly oriented towards selling goods in the world market. In recent decades, many countries have dealt with the world market situation through import substitution. In contrast, export orientation is a much more ambitious program. Using only import substitution a country can limit the kinds of goods that come into the country. With those limits it is not necessary to be as efficient as if they were trying to sell their own goods competitively on the international market. These three countries did at one time, and still do to some extent, protect their markets in various ways, but they have also gradually tried to liberalize, using guided programs of imports consistent with their programs of rapid development. However, their goal has always been to

INTRODUCTION

produce goods that are competitive on the world market. To achieve that required a strong commitment and it generated a pressure to understand world markets, as well as excitement and apprehension. This commitment to the world market and the sensitivity to what the world market required was a powerful driving force.

A sixth characteristic is that there has been a high level of development of human services and a rapid upgrading of skills. These countries started after World War II with an already high level of literacy. Japan had achieved virtual universal literacy by 1900. Korea and Taiwan also had very high levels of literacy partly as a result of Japanese occupation, but they also increased their investment in education after World War II. Indeed, large numbers of students go to middle school, higher school, and university in these countries. The proportion of students graduating from high school in Japan is about 94 percent compared to less than 80 percent in the United States. Taiwan and Korea are catching up rapidly. Both countries made a very heavy investment in skills training, but they also send a high proportion of their best graduates abroad to bring back the most advanced knowledge, especially from the U.S.

A seventh characteristic is that they have a disciplined and trained labor force. There is a paternalistic side to labor management, with efforts to provide good relations with the workers. But there is also a strict authoritarian side demanding strict discipline and commitment from their workers. Young people have been trained strictly in the school systems before they enter the work force. As students they have experienced the discipline of coming on time, performing certain tasks, and working hard to learn new tasks. They are therefore prepared to respond to the discipline that the large corporations required in the factories and more recently in the offices.

Eight, the populace has made temporary sacrifices in the

course of modernization to meet international markets. In order to make initial investments and to expand and meet the competition in the international markets, these countries have gone slowly in increasing personal consumption. The wage rates in the early days went up much more slowly than the rates of increase in production. Some Western economists complain that the Japanese have made many sacrifices and are still making many sacrifices. As somebody who has visited Japan regularly since 1958, I think that is nonsense. Since the 1950s the Japanese people have improved their standard of living far more than anybody in the West and, perhaps, in the world. Even Korea and Taiwan have improved rapidly. These states all first made early investments in industry before investment in state welfare. They were slow in responding to pollution problems. They were initially so oriented towards meeting goals for the international market that welfare, pollution, and salaries lagged behind the growth rate. Eventually when they began to turn their attention to these other areas, they began catching up very quickly.

Finally, the ninth characteristic is that these countries exhibit a very powerful drive to make it, to succeed. The drive I think has both an economic rationale and a political base. It also comes from a conscious effort by the elite in these countries to make people aware of their economic problems and to develop broader popular support.

The economic base stems from the fact that these countries have large populations for the size of their land area. They cannot support their population entirely by their own resources. Japan in the 1840s and 1850s was basically self-sufficient. Since that time its population grew so rapidly that already by the 1930s it did not have enough agricultural capability to support its own population. After World War II, Japan with six million returnees from overseas, and natural population growth, could not feed its population without exporting to pay for their imports. The Japanese are short not only of food but also of natural resources and energy. This urgency for exporting felt in Japan is also felt in South Korea and Taiwan. Because these countries must export in order to survive, they have a powerful, well organized push in that direction.

This powerful drive also has a strong political base in these countries. In Japan the political push is partially racial. It is a popular East-Asian belief that, "We orientals have been dominated by the white civilization and we are going to work very hard to catch up with them and overtake them." One should not underestimate the strength of that drive even today. In the case of South Korea the powerful political drive comes from feelings about Japan. South Koreans have expressed the belief that, "The Japanese have dominated and overwhelmed us, they occupied us so brutally for so many years that we are going to struggle and show the Japanese, and not let them dominate us any longer." Also of course Korea faces the political problem from the North. In the case of Taiwan it is perhaps less racial, but is related to fears about the mainland which until recently posed a serious military threat. More recently the threat is of isolation because of international pressures and of losing political recognition. In all these countries there is a very powerful economic and political-social push that has helped make these other things possible. I do not think any one or two or even three of these factors would be enough to explain why these countries have been so successful.

Because I spent 1987 in China and stayed in the southern part of China through the end of that year, I would like to make a few comments on the basic differences between these Asian industrializers and China. Then I would like to address the question of whether China might be able to make the same rapid breakthrough in industrial development.

CHINESE ECONOMIC POLICY

One of the big differences between China and these other places is that China is too large to benefit by relying primarily on export-like growth. Think about the difficulty the world, particularly the United States and Western Europe, has had in absorbing Japanese products. Then try to imagine that instead of 120 million Japanese, 1 billion, 100 million Chinese were producing equivalent amounts of goods. Imagine the difficulty the world market would have in absorbing those goods. There is no way the world can absorb enough goods to let the Chinese earn enough foreign currency per capita to achieve the level of success that smaller countries have achieved. Also, the biggest and most open economy in the world, the United States, is no longer as strong or as confident as it once was. When America begins to balance its budget and balance its trade it can not absorb as much import growth as when its markets were open and consumer purchases were growing rapidly.

Since China is so large, it will have great difficulty industrializing at the same pace throughout the whole country. Even Japan, Taiwan, and Korea experienced wide gaps in the early stages of their development. In China the gaps will be even greater. The more remote areas will undoubtedly have more difficulty. The more advanced areas within China will industrialize first. And these areas will begin to take on many of the characteristics of the rapidly growing East Asian countries.

China is different from these other East Asian countries in starting with a lower average income and a lower educational level than these countries did after World War II. Large problems of literacy remain. The average income is still low even by standards of Japan or Taiwan in the late 1940s or of Korea by the mid 1960s. Also because China started at an earlier stage with system of social security, the difficulty of producing goods for international markets is much greater. In China, cadres and state enterprise workers already enjoy an elaborate social welfare structure; while in other countries, the state came in much later with social welfare programs. It is thus difficult for China to be as lean and mean as these other countries in the early stages of making its goods competitive on the international markets.

Another of China's problems is its large nonmeritocratic bureaucracy. Most other countries had a fairly small elite meritocratic bureaucracy. In China the bureaucracy expanded rapidly after 1949, and for many years political criteria were critical. Therefore in the early years standards of intellectual ability and economic performance were not as important. Although this is rapidly changing, at this stage of development large numbers of cadres remain poorly educated by the standards of Japan, Taiwan, and South Korea.

Another characteristic results from the fact that as a socialist country, China has tried to run the economy itself. It did not originally leave economic management to private firms. In China as everywhere else in the socialist system, running the economy without the cooperation of private firms has proved unsuccessful. China is now trying to change its economic approach. The question is to what extent China can adopt the successful patterns of these other countries?

I am working this year in the province of Guangdong which is rapidly becoming a newly industrialized province. Within a decade, Guangdong with its 60 million people, especially the Hong Kong to Guangzhou corridor, may well look like one of East Asia's newly industrializing countries. Hong Kong is of course already vibrant, and its economy is becoming closely linked with the entire Pearl River delta area, including Guangzhou. This change is already spilling over into other parts of Guangdong. One could make a strong case that we are witnessing the beginning of the process of China joining the industrial age. Fellow China watchers are aware how difficult it is for even the best informed people to make accurate judgments about what is happening and developing in China, let alone make predictions about the future.

Let me suggest two or three reasons why I am optimistic that this part of China and probably other parts of China as well will make it. First is a powerful drive which comes partly from so much suffering during the Cultural Revolution. Officials in Guangdong have welcomed officials from Eastern Europe and Russia. These officials from other socialist countries have commented that they think reforms are likely to work better in China because of the inherent industriousness of the Chinese. While this drive is evident in China, there is a greater acceptance of the status quo in other socialist countries. However, in China the suffering of the Cultural Revolution, and the horrible awareness that so many lives were ruined, has become a catalyst for change. Another source of the drive comes from many overseas Chinese returning to China. These overseas returnees bring new ideas and material wealth which are hard for the Chinese to tolerate. When they see their Hong Kong friends and relatives with so many material things it is hard to say that they cannot have those things in China. Therefore Chinese determination to change and the willingness of overseas Chinese to provide new ideas, technology, capital, access and know-how for world markets is powerful. Also the Chinese are beginning to strive towards international markets.

China still faces many problems. There are indeed bureaucratic confusions, people with conservative attitudes threatened by change, and people who want to get rich quickly who don't realize how hard they must work. Some people want to acquire material goods by gift rather than by hard work. These problems are serious. I sense however, a tremendous determination and willingness to change and

INTRODUCTION

therefore I believe that in certain parts of China we will soon have the "Newly Industrialized Countries." They may not grow as fast. They have many disadvantages. But I believe they will be like "Newly Industrialized Countries" early in the 21st century.

ONE

DILEMMAS OF PARTIAL REFORM
STATE AND COLLECTIVE FIRMS VERSUS THE RURAL PRIVATE SECTOR[1]

David Zweig, Tufts University

To speed up the commercialization of China's rural economy, China's leaders decided to foster the growth of the rural private sector. The limited growth of the rural economy under the Maoist system of bureaucratic control and the improving rural standard of living following the liberalization and expansion of the private sector demonstrate the utility of this policy. Nevertheless, while studies addressing the re-emergence of the private sector in China have taken or will take as their starting point the myriad problems confronted by private entrepreneurs (Hershkovitz, 1985; Zweig, 1986), this paper looks at the dilemmas confronted by some of their competitors—the Supply and Marketing Cooperatives (S&M Coops), service companies and other state-run service industries.

This is no plaint plea for the superiority of socialism. However, approaching this reform from their perspective highlights the regulatory constraints that restrict the behavior of some rural collective and state firms even as the Maoist rural political economy is being deregulated and liberalized. These units are confronting reverse discrimination, a form of state-imposed, negative affirmative action, while other units, firms or businesses are receiving special dispensation or assistance. This approach also helps pinpoint one source of the lingering hostility to the rural reforms, and suggests that the level of opposition might drop if the reforms are deepened, rather than simply reversed; for these regulated units, simply stopping in midstream may be the worst of all worlds.

CHARACTERISTICS OF A PARTIAL REFORM

Under the Maoist rural political economy, resources flowed primarily through a vertical, bureaucratically dominated system which controlled the goods produced in the countryside and transferred them according to the dictates of a planned command economy. State and collective rural cadres from the county level down managed the labor, capital, and land of the collectivized peasant all the while withdrawing resources from the controlled flow of goods to expand their own sinecures (Zweig, 1989).

After nationwide introduction of the household responsibility system, China's reform leaders called for the second stage of the rural reforms. A market mechanism and increased privatization of the non-agricultural sector has increased the amount of goods flowing horizontally among individuals and units beyond the purview of the bureaucracy's vertical controls.[2] Also, to increase the free flow of commodities and services, the second stage of the rural reforms has introduced policies that weaken the state and collective sector's monopoly over rural marketing and services (Watson, 1988). In response, a resilient private sector now cuts into their markets and challenges them for business. And yet, as these commercial and service companies confront this new competition, they find themselves constrained by the "dilemmas of a partial reform."

Whereas one may think of a centrally planned economy as a highly regulated one—where guidelines emanating from the national government constrain most commercial interactions—a "partial reform" involves only limited deregulation and a limited introduction of market forces. First, the market remains constrained by continued intervention by state bureaucrats who seek to guide its development in ways that help them fulfill their political obligations. Second, these officials may also limit access to that market to protect those firms whose taxes are important revenue sources for the local government. Under these conditions only certain firms are free to operate according to market rules, while the decisions of others remain determined by the bureaucracy which limits their ability to compete in the market. And although these regulated units may even have a comparative advantage in the marketplace, under a partial reform local governments invoke certain rules or guidelines that impose some disadvantage on them.

To return to the vertical/horizontal perspective, under a partial reform some actors who were part of the vertical bureaucratic system are forced to continue to work under that system, and remain guided by its morality, while others are now free to work under a horizontal system, with fewer controls. And yet new opportunities which arise from the commercialization of the rural economy, the expansion of consumer demand for goods and services, new opportunities for privately transporting goods, the introduction of a dual price system—with goods sold outside the plan inviting the higher prices—and a variety of other semi-market opportunities, are expanding daily. These sectors are where much money can be made. Yet state and collective firms remain under the control of local governments who admonish them not to seek quick profits but direct them to demonstrate a socialist ethos when they "serve the people," while private businesses remain freer to serve the people for a quick profit.

Ironically, whereas under the Maoist system, when units controlling the vertical hierarchy utilized that power to dominate individual peasants and expand their resource base, during the current transitional stage of the reforms some of them, particularly commercial and service sector firms have tighter constraints placed on them than on private entrepreneurs who are better able to take advantage of the market opportunities now available under the reforms.

Finally, the continued existence of agricultural planning at both the national and local level forces state and local bureaucrats, who continue to receive planned quotas from above or who must ensure the meeting of societal needs, to respond in a manner which creates a new source of inequality unique to a partial reform. Driven to fulfill these responsibilities, they may impose production obligations on a particular set of units, such as those closer to a city or situated in valleys rather than hills, while leaving another set of units freer to pursue market-oriented interests. Moreover, these former units may be forced to sell these goods to the state at planned or negotiated prices, while the latter units are free to produce goods that can be sold at higher market prices. In this way, too, the partial nature of the reform places serious binds on one set of units while those in different locations or with different characteristics are free to make more money by participating in the newly expanding market sector.

DILEMMAS OF PARTIAL REFORM

Changes in the Rural Commercial System

The ongoing rural reforms, begun in 1983, but expanded in 1985, have increased the number of private businesses as well as their freedom to compete with collective and state companies. At the same time, these latter companies have remained under tight constraints, and have been pressured to reform their internal operations as well as their relationship to the external economic environment.

Growth and Problems of Private Businesses

Since 1978, the state has opted for "managing," rather than "policing," the private sector (Solinger, 1984:159), and has facilitated its development. As a result, between 1978 and mid-1986, the number of private businesses in towns, cities, and the rural areas, as well as their volume of sales, has increased dramatically (Table 1). Data from Nanjing City and Jiangpu County, near Nanjing, show a similar pattern of growth in the number of private businesses between 1978 and 1985 (See Table 2).

In Jiangpu County, many of these private businesses have focused on the food, service and commercial (store ownership) sector (Table 3), which compete directly with the Supply and Marketing Coops (hereafter S&M Coop) and the state-run service companies.

In this county in 1986, 1.4 million Renminbi (hereafter rmb) of a total business volume of 3.0 million rmb carried out by private businesses was earned in commercial exchange. Also, in Tangquan Township, within this county, of a total of 234 private businesses in 1984, 115 ran small shops, and another 31 were in the food business. In fact, according to local officials in Tangquan, the commercial sector developed too rapidly and quickly became overcrowded.

State policy has fostered this growth. Under State Council guidelines, private businesses, especially "those who indeed have difficulty in running a handicraft, repair

Table 1
DEVELOPMENT OF SMALL BUSINESS IN CHINA, 1981-1986
(in 1,000)

	1981	1982	1983	1984	1985	1986*
Number of Businesses						
nationwide	1,829	2,636	5,901	9,304	11,710	11,340
in cities and towns	868	1,132	1,706	2,222	2,798	2,734
in rural areas	961	1,504	4,195	7,082	8,915	8,610
compared with the preceding year (percentage change)	+44.1	+44.1	+124	+57.4	+25.9	—
Number of Persons						
nationwide	2,274	3,198	7,465	13,031	17,660	17,190
in cities and towns	1,056	1,358	2,086	2,911	3,839	3,790
in rural areas	1,281	1,840	5,378	10,120	13,822	13,400
compared with the preceding year		+40.6	+133	+74.3	+35.5	

*June

Source: *"Dangde shiyi jie sanzhong quanhui yilai geti gongshang hu jiben qingkuang"* (Basic data of individually owned small businesses since the Third Plenum of the Party's 11th Central Committee), *Gongshang xingzheng quanli* (Industry and Business Administration), no. 24, December 16, 1986, p. 13.

Table 2
DEVELOPMENT OF PRIVATE BUSINESS IN NANJING, 1978-1985
(in 1,000)

	1978	1980	1982	1983	1984	1985
Number of Households	1,738	5,535	10,443	21,577	34,137	49,977
Number Employed	1,758	5,770	10,788	24,191	42,465	61,634

Source: Nanjing Industrial and Commercial Management Bureau, June 3, 1986.

Table 3
INDIVIDUAL BUSINESS EMPLOYMENT STRUCTURE, JIANGPU COUNTY, 1978-1986

	1978	1980	1982	1983	1984	May 1986
Industry	22	27	71	106	286	341
Commerce	33	195	613	778	1,190	1,281
Food	41	57	108	137	384	369
Service	12	20	35	44	152	123
Repairs	17	53	91	104	249	252
Transport	-	0	2	2	275	1008
Construction	-	2	3	7	10	12
Other	3	3	4	5	5	20
Total						
Number of Households	128	357	927	1,183	2,551	3,406
Number of Employed	128	553	1,368	1,690	3,344	4,362

Source: Industrial and Commercial Bureau, Jiangpu County, June 10, 1986.

service, other service trade or catering service" were promised special treatment "in the form of loans, *favorable prices or tax reduction or exemption* or necessary technical assistance from the state" (State Council, 1984). When private businesses came under attack, Nanjing promulgated an eight point document in January 1985 on protecting the legal rights of the private businesses (*Baohu getihu de jingji hefa quanyi*). Yet many of these small businesses do not really need major tax breaks. In theory, taxes are assessed at 3-5 percent of sales plus a graduated tax ranging from 7 percent for a gross income of under 1,000 rmb/month to 60 percent on a gross of 30,000 rmb/month and above (Salem, 1987). According to owners of a private restaurant in Nanjing, they must pay 55 percent if they make over 6,000 rmb/year. But many private businesses do not keep good books or any books at all. According to one 1,000 household survey in Changchun in Jilin Province, 61 percent kept no books, 24 percent had records only of current accounts and a mere 15 percent had a full set of books (Salem, 1987). As a result, officials are often forced to estimate the income of private businesses. Local officials in Nanjing recognized that they really did not know how much the peasants made (Interview ICAMO). And if they remain as "specialized households," and not private businesses, they pay only income taxes, but not sales taxes.

To get more peasants into county towns to compete with extant collective businesses, the state introduced a new "long-term residence permit" (*chang zhu hukou*), which allows them to live in the towns, so long as they supply their own rice. Also, since a main problem for private businesses is a lack of suitable space for selling their wares, some counties, such as Jiangpu in Jiangsu Province and Shulu in Hebei Province (Blecher, 1987), are constructing buildings for them (*geti dalou*). Similarly, the Tangquan township government, a county-wide model for developing private businesses, charged 900 rmb and built them brightly painted stands along the main road. In other ways, too, township officials have demonstrated strong support for these businesses.[3]

Also, in Jiangpu County, the Individual Business Division (*getihu gu*) of the Commercial and Industrial Bureau (*gong shang ju*) was extremely supportive of the private businesses. These bureaucrats took their responsibility to help these private businesses seriously, and supported them in almost all conflicts with state or collective units. Although they establish guidelines and requirements that must be passed, such as tests for drivers and construction contractors, and health certificates for restaurateurs, they also give private businessmen classes to help them pass these tests; in most cases the teachers are state cadres.

Still, private businesses have confronted serious problems since 1978 (Solinger, 1984:202-203). S&M Coops and other state and collective enterprises have expended great efforts to undermine the growth of the private businesses. They have approached the Commercial and Industrial Bureau officials to get them to revoke the licenses they had given to the private businesses.[4] Overall, the kinds of restrictions and hostile reactions they have met are legion.[5] Getting supplies is also a major problem, although deregulating the markets for timber, grain and other products in 1985 alleviated some of these difficulties. For example, many of them need coal for cooking, but firms established after July 1, 1985, cannot buy coal from the state plan and thus pay higher coal costs (Interview FCD).

Government constraints on marketing and wholesaling leave private firms dependent on their competitors, the S&M Coops. According to some sources, private firms cannot go into the wholesale business very easily, reportedly because the government still sees the state and collective as the major marketers (Interview NCIB). As a result, private businesses in more distant rural areas must buy from their competition, the S&M Coops, which will not supply them with top quality goods (Interview LYL; *NMB*, October 19, 1985). Even increases in the amount of agricultural and industrial products have not totally undermined the power the S&M Coops wielded when supplies were short; S&M Coops remain a critical channel for peasants to market some of their surplus.

According to the directors of one suburban Nanjing S&M Coop, the private businesses with whom they compete are less under the control of the S&M Coops than those businesses in more distant locations. In the near suburbs, individual businesses buy about 65 percent of their goods from the S&M Coop, but they compete on over 95 percent of the goods the township sells. However, "in the more distant countryside,...they must rely on the S&M Coop more to buy their goods. This gives the S&M Coop in those areas more power" (Interview ZTSSMC).

CHANGES FOR THE STATE/COLLECTIVE UNITS

The ownership of S&M Coops has shifted back and forth from cooperative (1949-1958, 1962-1970, 1975-1978) to state (1958-1962, 1970-1975, 1978-1982) ownership (Han, 1987). Thus, in February 1982, the S&M Coops' administrative organization was incorporated into the Commerce Ministry to show that the S&M Coop system was not a separate, nationalized commercial network (Zhu, 1986). In response, Jiangsu Province in 1983 changed all S&M Coops from "all people's ownership" to collective ownership. The state no longer covers their losses, particularly after Central Document No. 1 of 1985 announced that all S&M Coops "should have completely independent accounting systems, be responsible for their own profits and losses, and operate on their own under democratic operation by the masses" (C.D., 1985). Also, all new employees with urban residence permits would be collective, not state, workers and S&M Coops could now hire peasant "contract workers" (*hetong gong*).

To replace the loss of state funding, S&M Coops can now raise funds directly from the peasants and the results have been significant.[6] In return, the Coops now return dividends to the peasants at year's end. Also, since 1984, over 77 percent of the S&M Coops in Jiangsu had divided their various divisions, such as food companies and individual stores, into 9490 financially independent units (Jiangsu, 1985). Yet, following the cancellation of compulsory sales of agricultural products in 1985, S&M Coops are now expected to help peasants market their produce all over the country (*NMB*, August 31, 1985:2), expand the information network for marketing and technical innovation (*NMB*, August 1,

1985:2), make loans to peasant entrepreneurs (*NMB*, November 7, 1985:2) and, in particular, promote the commercialization of agriculture (Fuyang, 1984). Many companies are clearly reluctant to perform these kinds of services.

Thus when I went to China to do research in summer 1986, I expected to find the villain among these semi-state, semi-collective enterprises. But interviews in rural Nanjing showed that at this critical stage in the rural reforms, governmental organizations at both the city and county level still control major parts of the rural economy. To maintain social and economic order, they are demanding that many of these units follow the plan or "serve the people." At the same time, they are responding to signals from the national level to loosen the constraints on the private sector, as well as on other parts of the collective sector, creating serious disadvantages for units still working under local bureaucratic control. In the end, I found a partially planned, partially market-oriented economy, with local officials still orchestrating a great deal of local economic activity, leaving some sectors better able to compete under these different types of markets than others.

DILEMMAS FOR STATE/COLLECTIVE FIRMS UNDER THE REFORMS

According to the *Beijing Review*, the S&M Coops are the main target of the second stage of rural reforms (*BR*, May 11, 1987:17), while Central Document No. 1 (1985), which deregulated the exchange of most rural commodities, called on state companies to "enliven their operation because they will be subject to market regulation." At the same time, the state has called on them to help peasants prosper, even though peasant private businesses may compete with shops or processing factories now run by S&M Coops and other state-run businesses (*NMB*, August 1, 1985:2), such as food wholesaling and transport.

Thus S&M Coops, service companies—which run hotels, restaurants, bathhouses, barbershops and photo shops in county towns—and other similar firms, such as "small collectives" (*xiao jiti* or *he zuo qiye*) in township towns,[7] face major problems resulting from the state's financial and tax policy, local labor regulations, heavy retirement demands, and local political interference.

PROBLEMS OF FINANCE AND TAXES

According to Chinese sources, S&M Coops in many ways are much worse off than state firms, even though they are free to go out and raise money (Meng, 1986). First, the Finance Ministry regulates their expenditures while taxing some of the bonuses for their employees. Second, profits on agricultural products are small, yet they get no credit assistance to cover the seasonal outlay of funds, as do state grain companies. Third, their rate of taxation rose from 39 to 50 percent, the same as state companies, but unlike state wholesalers who do not pay taxes, Coop wholesalers pay a 10 percent tax. Finally, according to Meng, even though their financial resources are strained, they remain the best source of funds for local governments and financial institutions who increase the contributions demanded of them.

These kinds of financial problems were raised by officials at the lower levels. When a county S&M Coop manager compared his problems to those of the collective enterprises, he complained that if they open up new stores, they get only one year tax free, the same as state companies, whereas if the collective opens a store or a factory, they get three years tax free. They also face stronger restrictions on the bonuses they give out, a point also raised by Meng's article. If the bonuses they give out are more than two and a half times the monthly salary, they must pay taxes on that money.[8]

Finally, due to bureaucratic complications, all transactions of the S&M Coop must go through the bank, so it takes them several more days to complete transactions than private businessmen who often have no bank accounts. According to another source, as of 1984-1985, cooperative stores (*hezuo zongdian*) in local towns could draw out only 30 rmb of their funds from the bank without permission, while managers of smaller branch shops (*menshi bu*) could draw out only 10 rmb (Small Towns, 1986). Also, state service companies work under regulations which stipulate the number of people needed to run a restaurant—a chef, a server, someone to take in the money and someone to buy the food. This imposed division of labor increases their labor costs as compared to private restaurants who do not confront such regulations.

PROBLEMS OF LABOR ALLOCATION

Due to recent policy changes, these firms could now legally hire contract laborers; however, the majority of workers in Jiangpu and Nanjing S&M Coops who were hired under previous policies are still collective or state workers. As of 1986, only 20 percent in Jiangpu county's S&M Coops, and only three percent (10 of 300) in Zijingshan Township's S&M Coop were contract laborers. The greatest impediment comes from the Labor Department (*renshi bumen*); as an official in Jiangpu complained, the reality of hiring is much more complicated than the "spirit of the documents."

While private businesses are free to hire and fire contract workers, Tangquan Township's S&M Coop cannot. The problem originates in the large number of unemployed youths in rural towns who have urban residence permits, but who refuse to work in unstable township factories. The Labor Bureau, which must find these urban residents jobs, has turned to the S&M Coops for help, demanding that they hire the children of Coop workers first. But if the Coop is to improve its level of service and finances, it needs to be able to fire slovenly workers;[9] yet the children of Coop workers, as urban residents, must be hired as semi-permanent collective workers, who short of committing major crimes, cannot be fired. From 1983 to 1985, S&M Coops in Jiangpu needed only the permission of the county government to hire peasant workers, but by late 1985 the Labor Bureau had reasserted its control and told the S&M Coop that if they hired peasants, the Labor Bureau would not help Coop employees' children find jobs in the state sector. Similarly, in Zijingshan Township, outside Nanjing, the S&M Coop has been unable to introduce a responsibility system for each worker because the workers are resisting it.

The county service company works under similar constraints. Before 1982, if it could not make enough money to pay its workers, it could bargain with the state for a cut in its income tax. But now, each store is an independent enterprise, confronting a hard budget constraint, so if it does poorly, its workers salaries are cut. They, too, do not control their own hiring and firing. Since 1980, Jiangpu's Labor Bureau has given them 20 new workers, and whenever one retires, the state sends a replacement, whether they want one or not.

RETIREMENT PROBLEMS

A major problem for all these firms is their responsibility for supplying retirement benefits to their former workers, even though for most of the past 20 years their profits went to the state's coffers not into the S&M Coop system. But now, as older workers retire, the S&M system is expected to cover their retirement benefits from its management costs. In some places, one current employee may support .5 to 2 retired workers, undermining current workers' benefits and labor enthusiasm (Meng, 1986).

In Jiangpu's service company, retirement benefits were a major problem. People who retired after 20 years were receiving 75 percent of their salary, while those who worked 15-20 years were getting 70 percent of their salary; all of them received free medical treatment. These funds reportedly came from this company.

In Tangquan Township, the "small collective"—where 60 retired workers are supported by 70 current employees—pays 3,000 rmb/month in salaries and medicine for retired workers. They, too, are an independent accounting unit, so poor business means lower salaries. A 1986 local recession meant that 82 workers in 1985, with a profit of 35,000 rmb, became only 68 current employees in 1986 and profits of only 15,000 rmb. Six workers retired, but others quit because of unstable salaries. Officials believed that private businesses took away 40 percent of their profits. Still, they had to continue to support their retired workers.

Finally, when a state or collective-run shop goes broke, it must take care of its workers (Interview LYL, June 26, 1986), while peasants who go bankrupt can go back to the fields.

POLITICAL PROBLEMS

S&M Coops confront a great deal of local interference, even as they try to balance their books. Although they are the same as other "large collectives" (*da jiti*), they are supposed to "serve the people," rather than be "too busy managing their own businesses, making money, issuing rewards or even competing for profits...struggling to handle that which yields big profits while leaving those making little profits alone, thus lacking sufficient consideration for their social effects" (*Economic Daily*, 1985). So that peasants will not scream about their difficulties marketing their produce, the county government forces the Coops to buy products they do not want. Yet if the Coop loses money on these deals, the state does not supplement their funds. Similarly, a Coop plan to combine several unprofitable stores in neighboring townships was turned down by the government because it could inconvenience the peasants. Thus although these firms are expected to stand on their own financially (*zi fu ying kui*), they are not totally free to pursue that goal.

Also some county officials may favor private businesses. Private businesses serve some needs of the general population better than the state and collective companies; also local governments may feel political pressure to conform to public policy and demonstrate that the private sector is growing under their stewardship. As a result, in Jiangpu County, there was a close relationship between the Private Business Association (*getihu xiehui*) and the Industrial and Commercial Bureau; in fact, they are run out of the same building—their signs hang side by side on the building facing the street—and the deputy head of the ICB, a former director of the ICB, heads the association. As a result, managers of collective businesses complained that the state checks the quality of produce in their shops more carefully and more often than that of the private shops. In fact, one township official in the S&M system argued that each time the ICB, the Tax Office (*shuiwu ju*), the Price Investigation Office (*wujia jiancha suo*) or the Health Bureau came to investigate, they only checked those shops run by the S&M Coop. According to him, "we have raised this many times, and only now is there beginning to be a change." In fact, while I was in the township, the restaurant run by the S&M Coop was closed temporarily for health violations.

Many of these local problems result from a lack of political support at higher levels of the political system. Nationally the S&M Coop organization is an "empty shell" (Meng, 1986), and is part of the Ministry of Commerce, so their fate under the reforms depends on that ministry's political clout. They are particularly vulnerable on issues of price reduction, which often harm their income, especially in localities where there is a high degree of specialization (Meng, 1986). Also, at the city level they are under the Commerce Department, so although they compete in the county seat with stores run by the Commerce Bureau, they buy their own products from

Commerce Bureau wholesalers who are likely to save the best products for stores in their own system.

Nevertheless, officials in Nanjing's Rural Work Department felt that the township S&M Coop had much more freedom today than under the "radical line," when the commune party committee's horizontal (*kuai kuai lingdao*) control gave commune leaders great influence over the grain station, the S&M Coop, the bank, etc. Now a major part of the reform is to get the township government to reduce its efforts to control local coop stores. According to Rural Work Department officials, many township S&M Coops do not listen to the township government, preferring instead to respond to the vertical authority (*tiao tiao lingdao*) of their county bosses, "because if they go bankrupt, the upper levels are responsible" (Interview NRWD, May 30, 1986). Still, if the ICB and the Individual Business Association are so closely tied nationwide, competitors of these private businesses may face continued discrimination.

COMMERCIALIZATION OF THE RURAL ECONOMY

Finally, the continued existence of grain quotas plus local governmental responsibilities for ensuring urban food supplies, both reflections of this dilemma of a limited reform, demonstrate how concomitant existence of plan and market generates new inequalities of opportunity. In both cases, units that do not fall under the constraints of the plan and are free to pursue other economic opportunities have profited greatly from commercialization; those restrained by the plan have done less well. At the same time, state fish companies have been unable to compete with peasant fish farms and most face serious economic difficulties.

Although the state formally ended the compulsory grain sales (Oi, 1986), each county, township, and village still receives a production quota. However, because land can now also be used for commercial endeavors, decisions on its use are much more important than before 1978 when marketing opportunities were minimal. Thus units that avoid grain quotas can better profit from new opportunities, creating inequalities among villages or brigades in the same township.

In Tangquan Township, hilly brigades which quickly began to raise tree seedlings got very wealthy in 1983-1985. This occurred in part because the township lowered their grain quota. However, crops grown by paddy areas are more tightly regulated; only recently have they shifted some of their poorest land away from grain, expanding their melon production and turning some of the land into fish ponds. As a result, their living standard is visibly lower than hilly teams.

Tangquan Township's overall prosperity has resulted in part from a low grain quota which other townships supplement. Nanjing officials recognized their dilemma. "No doubt, we need to adjust the grain quota from the county level and below to allow those areas that specialize to pursue those interests, while at the same time ensure the grain output. But if the other area helps you out by helping to fulfill your quota, then you need to pay them back" (Interview NRWD, May 30, 1986). At this point in the reforms, Tangquan's limited quota helped them grow rich at the expense of other localities who are fulfilling their quota. The fairest solution would be either to force everyone to grow grain or simply let the market come into full play.

The production of food—such as vegetables and milk—was still regulated in parts of rural China in 1986 as well. In fact, the policy of the "cities leading the counties" (*shi lingdao xian*) was instituted to help city governments control the crops grown for urban consumption. To ensure sufficient supply, 80 percent of the crops grown in a suburban Nanjing township were set by the city and only 20 percent were planted at their own discretion; if they ignored the plan and missed their vegetable quota, the state would cut their grain ration, forcing them to buy expensive grain on the free market (Interview MQZ, June 3, 1986). The city also relied on its party authority to ensure cooperation with its directives.

In fact, in all conversations top party officials in this township constantly stressed their role in feeding the city; clearly the city officials had made their point emphatically to this township's leaders. But in private conversations other township officials expressed disappointment and frustration that some of their leaders were too concerned with fulfilling the state plan. They had devised some plans to take advantage of the commercialization of the rural economy, but to this point their suggestions had been turned down.

As a result of these and other constraints, this suburban township had not industrialized as rapidly as neighboring townships further away from the city because it had to use its funds to ensure a rich supply of good quality vegetables. Also, because this township must sell all its milk to the state, they cannot establish their own milk processing factory and process their own milk. Were they not part of this plan, they might find it profitable to process and sell their own milk in the city. But the city, which runs its own milk factory, refused their request to build a factory, thereby preventing competition.

Finally, a freer fish market, where peasant entrepreneurs are selling directly to factory mess halls, has led to hard times for state-run fish companies in Nanjing. Their share of the market has shrunk considerably. In 1985, almost one million kg. of fish were sold that way in Nanjing (Interview NRWD, May 28, 1986). Moreover, state company sales of 20 million kg. in 1983 dropped to 9.5 million kg. in 1984. As of 1985, when the state stopped giving them a supplement, they had problems feeding their workers. These problems arose "because the state controls on them were quite tight, with restrictions that others did not have" (Interview NRWD, May 28, 1986). While one would need to analyze this problem more meticulously, similar problems have developed for numerous other food companies in Nanjing (Table 4), suggesting that this is a common dilemma at this stage of the reforms.

Table 4
PRIVATE SALES AS A PERCENTAGE OF STATE AGRICULTURAL TRADE, NANJING, 1983 AND 1985

Products	1983	As %age of State sales	1985	As %age of State sales
Grain	8,735	2.1	3,610	0.8**
Pork	419	.84	2,234	4.9
Beef	152	11.53	293	24.2
Eggs	3,151	34.78	4,430	67.8
Fowl	3,380	41.60	5,040	133.1
Fish	2,847	156.8	5,800	756.0
Vegetables	36,712	16.46	68,890	42.7

** people are eating less grain in Nanjing now and more meat, fish, and vegetables.
Source: Industrial and Commercial Management Bureau, Nanjing, June 3, 1986.

REFLECTIONS ON A VISIT TWO YEARS LATER

This paper was based on data collected in summer 1986 and was written in summer 1987. Since that time I have returned very briefly to Jiangpu County and have had the opportunity to talk with some of the same actors. Rather than rewrite this paper in its entirety—an unnecessary task given that this paper recognized the transitional nature of the phenomena it described—I would simply add some comments describing the major changes I observed.

First, continued reforms have relieved some hostility generated by constraints imposed by local governments. For example, according to the new director of the S&M Coop in Jiangpu, their economic situation has improved dramatically since 1986 due to an expansion of their wholesale efforts. Whereas in 1986 their sales and purchases occurred only within the county, they are now able to buy better quality products outside the county and market them in other counties as well. One of their stores in a township bordering Anhui Province serves as a wholesale outlet for sales in Anhui. As a result, sales in 1986 of over 90 million rmb have increased by 55 percent (not controlling for inflation); by June 1988, their sales were already over 70 million rmb.

Labor and retirement problems have receded as well. According to this director, pressures from the Labor Bureau to hire children of Coop workers have passed and reflected a particular period when jobs in the county were few and unemployment high. Perhaps, too, rural industries have expanded and become slightly more stable. As for pension problems, efforts are being carried out nationwide to resolve this welfare dilemma. In Jiangpu, the S&M Coop now draws out 23 percent of total salaries of all employees from its yearly profits, and gives it to a state labor insurance company. The insurance company handles all retirement payments. While one wonders what happens to companies who cannot afford these payments, under this new system retired workers remain much more secure since their retirement payments do not come directly from their old firms.

Finally, the recent expansion of the S&M Coop in Jiangpu may merely reflect the slow nature of commercial reform that had taken place in this county. Since 1986, however, several older officials have been replaced by more active, younger county officials who are seeking to increase the pace of development. One need only compare Jiangpu with the more prosperous and powerful S&M Coop organizations in southern Jiangsu Province which rapidly expanded their commercial, wholesale, and industrial activities. For example, the S&M Coop in Wujiang County, outside Suzhou City, which I visited in summer 1988, now owns 42 factories and employs 8000 workers, which according to the director is the result of their open-minded attitude towards business opportunities. Thus in 1984 they expanded their commercial and industrial activity. Under a freer economy, the S&M Coop in Jiangpu County may be also able to grow much more rapidly as well.

CONCLUSION

The dilemmas described here, which result from continued regulatory controls on some sectors of the rural economy combined with deregulation of other sectors or actors, are part of the distortions that develop in socialist systems that are in the process of change. Nevertheless, they do create inequitable conditions for some firms, even while others receive special advantages. Ironically, the firms that are suffering most are those who held monopolistic control over the commercial sector during the Maoist era, and who tried to undermine the reforms in their early stages. Similarly, some units who are losing out on new economic opportunities had received advantages under the planned economy, since they were allowed to grow vegetables, while others were kept in poverty by irrational demands of a state that enforced high levels of grain production.

Even though one major goal of the reform is to get politics out of the economy, the partial nature of the reform has ensured that politics remain an important determinant of one's position in the marketplace. The price of remaining under state control can be high, making opportunities to get out from under state regulations dependent on political relationships. Politics also determines which units remain under local political control. From the county level down, local governments still intervene in the market and force firms to follow the government's political agenda rather than the dictates of their own economic logic.

China's economists and social scientists recognize these dilemmas, but the problem is what to do next? Should rural China push forward and further deregulate the rural economy? If they did, would the state still be able to ensure that there would be enough grain for urban residents? As it is, the state is pressed to match grain production to population growth. In a non-regulated economy, larger firms are likely to do well, allowing collective and state firms to compete better with the private businesses; but if their size and power dooms the rural private sector to oblivion before it establishes itself firmly, services in the rural areas would suffer again.

Yet are continued regulations and discriminatory policies not themselves somewhat self-defeating due to the hostility they generate against the reforms? Data suggest that even as these firms try to reform and compete, continuing regulations and pro-private sector, affirmative action policies create a great deal of local tension. Moreover, under the current situation, these firms are having difficulty reforming internally and competing externally. But can the state really allow them to go bankrupt, without being certain that the private sector will fill in these employment gaps?

Clearly, these are not easy questions for the Chinese to answer. Western economic experience shows that there are both benefits and costs in deregulating an industry. However, China now finds itself in a rather unique situation, with both regulated and non-regulated firms competing in the same sector. Future solutions will come only from the kind of ongoing rural research in which the Chinese are currently involved. Still, in any reform, there are many dilemmas but few easy solutions.

NOTES

1. Research for this paper was funded by a Research Grant from the Social Sciences and Humanities Research Council of Canada in Ottawa and the Sackler Foundation in Washington, DC. Research in China was arranged by Nanjing University's Foreign Affairs Office and the Department of Economics, and was conducted in association with Fan Conglai of the Department of Economics. Research assistance was provided by Sonny Lo at the University of Waterloo, Merrit Maddux at The Fletcher School of Law and Diplomacy, Tufts University, and Nancy Hearst at the John King Fairbank Center for East Asian Research, Harvard University. Thanks to Louis Putterman, Kathleen Hartford, Jean C. Oi, Dwight Perkins, and especially Bruce L. Reynolds for editorial comments.
2. Yet while access to rural resources has diminished for many rural bureaucrats, some organizations, such as brigade party branches and township (i.e., administrative) organizations have maintained significant control over rural resources and thus have lost little political power (Oi, 1986). Township enterprises run by these administrative organs have boomed as well. Finally, some very basic rural cadres have abandoned the collectives and struck out on their own as entrepreneurs (Burns, 1986).
3. Although state policy prohibits cadres from running private businesses, a school teacher opened a small factory employing over 20 workers, but used his 20-year-old daughter as the formal owner. Even though everyone knew this was a cover—in fact, some people wrote letters of protest—township officials responsible for private businesses turned a blind eye to this infraction. Also, these officials helped the families selling pork in the free market to buy a freezer.
4. A rural Nanjing informant admitted that a director of a township S&M Coop had tried to persuade the local CIB official to restrict private business activities. These kinds of reports appeared frequently in 1983 in *Peasant Daily* (*Nongmin Bao*).
5. For a more in-depth discussion of the problems they face see Hershkovitz, 1985 and Yudkin, 1986.
6. By the end of 1984, peasants had invested over 57 million rmb in

Coops in Jiangsu, increasing the liquid capital from 4.79 percent to 11.64 percent. See Jiangsu, 1985. In Qingdao's six counties and one suburban district, capital under the control of the S&M Coop coming from the peasants' investment rose from two to three percent of total capital in 1984 to 20 percent in 1986 (*BR*, May 11, 1987:18).
7. In many parts of rural China, these "small collectives" were made up of privately owned stores that were collectivized in the 1950s. They ran many of the stores in the villages and brigades during the collective era, but today are supervised by the S&M Coop system.
8. Nevertheless, one can understand the state's efforts to stop these firms from decreasing their pre-tax profits, and therefore their taxable income, through major bonus disbursements.
9. At the same time, while it recognizes the problems in carrying out these kinds of internal reforms, the press lauds those S&M Coops that are able to introduce these "systems of responsibility." See "Take Responsibility for Managing the Wholesale Department and the Retail Shop of the S&M Coop," in *Peasant Daily*, September 12, 1985:2.

BIBLIOGRAPHY

All references to interviews followed by a series of letters refer to interviews carried out in summer 1986. Copies of those interviews are available on request.

Blecher, 1987. Personal communication with the author.

BR, 1987. Han Baocheng. "Farmers Active in Commercial Sector." *Beijing Review* 19 (May 11, 1987), p. 17.

Burns, 1986. John P. Burns. "Local Cadre Accommodation to the Responsibility System in Rural China." *Pacific Affairs* 58:4 (Winter 1985-1986).

CD, 1985. "Ten Policies of the CPC Central Committee and the State Council for further invigorating the rural economy (January 1, 1985)." *Foreign Broadcast Information Service* (March 25, 1985), pp. K1-7.

Economic Daily, 1985. *Economic Daily (Jingji ribao)*. December 5, 1985, in *Joint Publication Research Service*-CAG-86-027. July 14, 1986.

Fuyang, 1984. "Fuyang County Supply and Marketing Coop Enthusiastically Serves the Development of Commercialized Agricultural Production." *Develop the Rural Commodity Economy (Fazhan nongcun shangpin jingji)*. Agricultural Publishing House, 1984, pp. 268-278.

Hershkovitz, 1985. Linda Hershkovitz. "The Fruits of Ambivalence: China's Urban Individual Economy." *Pacific Affairs* 58:3 (Fall 1985), pp. 427-450.

Jiangsu, 1985. *The Situation of Economic and Social Development in Jiangsu, 1984 (Jiangsu jingji he shehui fazhan gaikuang, 1984)*. Jiangsu People's Publication House, 1985.

Meng, 1986. Meng Lijia. "Discussion of the Reform of the Structure of the Supply and Marketing Cooperatives." *Nongye jingji wenti* (Problems in Agricural Economics) 2 (1986), pp. 19-21.

NMB. *Nongmin bao* (Peasant News).

Oi, 1986. Jean C. Oi. "Peasant Grain Marketing and State Procurement: China's Grain Contracting System." *The China Quarterly* 106 (June 1986), pp. 272-290.

Oi, 1986a. Jean C. Oi. "Commercializing China's Rural Cadres." *Problems of Communism* 25 (September-October 1986), pp. 1-15.

Salem, 1987. Ellen Salem. "Peddling the Private Road." *Far Eastern Economic Review* 8 (October 1987), p. 106.

Small Towns, 1986. Jiangsu Province Small Town Research Group. "Impact of Urban and Rural Reforms on Small Town and City Development." *Xiao chengzhen, xin kai tuo* (Small Towns, A New Beginning). Huaiyang, Jiangsu People's Publishing House, 1986.

Solinger, 1984. Dorothy J. Solinger. *Chinese Business Under Socialism*. Berkeley. University of California Press, 1984.

State Council, 1984. "Certain Regulations Governing Individual Industry and Commerce in Rural Areas." February 27, 1984. In *Foreign Broadcast Information Service*. March 13, 1984.

Watson, 1988. "The Reform of Agricultural Marketing in China Since 1978." *The China Quarterly* 113 (March 1988), pp. 1-28.

Yudkin, 1986. Marcia Yudkin. *Making Good: Private Business in Socialist China*. Beijing. Foreign Languages Press, 1986.

Zhu, 1986. Zhu Weiwen. "Gongxiao hezuoshe de xingzhi jiqi gaige" (The nature of the supply and marketing coops and their reform). *Nongye jingji wenti* (Problems in Agricultural Economics) 2 (1986), pp. 14-16.

Zweig, 1986. David Zweig. "Prosperity and Conflict in Rural China." *The China Quarterly* 105 (March 1986), pp. 1-18.

Zweig, 1989. David Zweig. *Agrarian Radicalism in China, 1968-1981*. Cambridge Massachusetts. Harvard University Press, 1989.

TWO

STILL ON THE COLLECTIVE ROAD
LIMITED REFORM IN A NORTH CHINA VILLAGE[1]

Thomas B. Gold, University of California–Berkeley

From the reports of some exuberant foreign and Chinese observers, one might suppose that the time to prepare burial clothes for China's collective agriculture had arrived. "Decollectivization" in the two senses of dismantling the commune-brigade-team system and the supplanting of collective endeavors by private production would seem to describe accurately current rural reality. Furthermore, one might conclude that rural cadres, with their former responsibilities removed, have hopped on the wind of going it alone, either starting their own private enterprises benefiting from extensive *guanxi* networks woven from years of bureaucratic experience, or fattening off the riches accumulated by others via corruption and extortion.

My own travels in southeast China, in particular the Minnan region of Fujian, left a deep impression of peasant entrepreneurship run amok. Peasants had apparently deserted the fields for workshops, stores, food stalls, vans, trucks or whatever investment avenues they discovered. With their cash incomes they built imposing two-story cement houses seemingly wherever they wished. Only the dimwitted still labored in the fields.

It was with impressions along these lines that I embarked on three weeks of field work in Fengjia Village, Zouping County, Shandong, in July 1986. I departed having learned that, not surprisingly, there are many routes to decollectivization—to say nothing of fast and slow lanes, and authorities on patrol.

My research objective was to gauge the extent of peasant entrepreneurship: to what degree had peasants been structurally liberated from the collective, planned economy and how much of the new freedom to maneuver had they directed to private enterprises, risking their own capital? The brief answer, for Fengjia Village, at least, is that the peasants had hardly been released from the collective planned structure at all and that almost none of their savings were being channelled to entrepreneurial activity.

In this paper, I will describe the manner in which the village leadership has continued to maintain the power of the collective over the peasants and the nature of the "entrepreneurship" that has emerged. I look first at agricultural activity, then at industrial and commercial undertakings.

THE RURAL REFORMS

The Chinese Communist Party's (CCP) bold reform of the commune system it had labored for two decades to build has drawn worldwide attention and emulation. The 1980s have witnessed a trend toward privatization in socialist and capitalist, developed and underdeveloped economies alike. The CCP's across-the-board criticism of both its Soviet and Maoist experiences and its declaration of embarking on uncharted waters in search of "socialism with Chinese characteristics" have been a dynamic part of this trend. China's resultant rapid economic growth has had global ramifications. Socialist developing nations like Vietnam and even the USSR itself, have begun to reconceptualize socialism as both an economic and sociopolitical system.

China's rural reforms have been of particular interest. Heretofore, existing socialist nations have been notorious for their brutal collectivization, emphasis on heavy industry at the expense of agriculture, and subsequent inability to feed themselves. In spite of heading a peasant-based revolution and repeatedly affirming the necessity of supporting the rural sector, the CCP also discriminated against the rural sector and failed to sustain significant improvements in agricultural production or the peasants' standard of living after the initial land reform. When the CCP's reformist leadership redirected the focus of work away from class struggle to economic development in December 1978, in an attempt to reestablish its legitimacy, it selected agriculture as the first of the Four Modernizations.

A number of scholars have analyzed China's Agricultural Responsibility System (ARS),[2] so I will not review it except to emphasize that its essence is the replacement of collective, commandist, planned production and compulsory delivery with contracts signed by households, individuals, or groups to sell a portion of output to the state for cash, and the opening of free markets for the disposal of the rest. The three level commune-brigade-team structure has been officially renamed (*xiang*, village, sub-village), and the power of cadres to meddle in the economy has been curtailed, at least on paper.

FENGJIA VILLAGE—AN INTRODUCTION

Fengjia Village is one of 40 natural villages[3] (*cun*) in Sunzhen *xiang*, formerly Sunzhen commune. Sunzhen (1986, population 35,951) is one of 14 *xiang* and three *zhen* in Zouping County (1986, population 645,000) which lies northeast of the Shandong provincial capital, Jinan. Of Zouping's 144,446 households (*hu*) in 1986, 142,060 were classified as peasant, meaning that they ate primarily food they grew themselves, although many now engage in industrial and commercial sideline activities. Most of the county's 1,250 square kilometers lie in flatland and 1,200,000 *mu* are arable. The main crops are winter wheat and corn. Prior to the Third Plenum of the Eleventh Central Committee in December, 1978, Zouping concentrated on grain production. In 1978, agricultural output value was 103,040,000 rmb. Industry was 50,300,000 rmb. Since then, the leadership has strongly encouraged industry and, by 1985, industrial production value reached 247,700,000 rmb to agriculture's 252,080,000 rmb. In the 1930s, Zouping was the site of Liang Shuming's Rural Reconstruction Project, so it has a legacy of officially sponsored economic development and a respect for education.[4]

Sunzhen's area is slightly more than 100 sq. km. with 94,566 *mu* of arable land. Its major crops are cotton, wheat and corn. Per capita income for 1985 was 369 rmb. The *xiang* government was officially established in May 1984.

Fengjia Brigade became Fengjia Village (*Fengjiacun*) in November 1984, although it is still commonly referred to as "Fengjia Brigade." It lies two kilometers west of Sunzhen town, easily accessed by a recently-paved road.

Feng Yongxi has been party secretary since 1961. The five-member elected village council (*cunmin weiyuanhui*) was established relatively late, in March 1985. Feng Yongxi is not on the council, but party and village offices are in the same building. Two of the five members had been team leaders, and the others served in the brigade leadership. One is a woman.

Fengjia Village has 2,600 *mu* of arable land, planted primarily in winter wheat, corn, cotton and peanuts. As of August 1986, there were 273 households comprising 1,120 people, 60 percent of whom are surnamed Feng. Thirty people are members of the CCP, 20 of the Communist Youth League. The village is subdivided into three village groups (*cunmin xiaozu*), the same as the number of teams in the old system. They are not geographically identical because, between 1981 and 1983, all of the old houses were demolished and the residents moved into new collective buildings. The village had constructed 480 *jian* (the space of a door or window) by mid-1986. Households bought anywhere from four to 12 *jian* of a main dwelling with southern exposure in a courtyard that also includes a kitchen, pigpen-*cum*-toilet, and storage facility. Most households own five to seven *jian*. They pay 720 rmb per *jian* of brick main dwelling and 350 per *jian* for the other buildings. Although there is an interest-free installment plan, most pay in one lump sum.

Furnishing was stupefyingly conformist. Against the wall opposite the door in the main room sits a sturdy wooden table and chairs and a mantle. Commonly arrayed on the mantle are a radio, tape recorder, fan, plastic flowers in a vase, and souvenirs or handicrafts. A landscape painting commonly broken into four scrolls hangs above the table. Two calligraphy scrolls flank the landscape. For older people, the scrolls are painted; for younger couples, they are frequently mass-produced gaudily colored plastic. This fixture is a gift from the brigade-village. The room also has a large *kang*, a TV set and chairs. The chief observable difference in interior decor between younger and older couples, other than the number of plastic knickknacks, was that the former glued pinups of movie stars and athletes on the walls.

In 1985, per capita income was 954.30 rmb, including the value of contracted grain, cash wages and sidelines.[5] Based on a survey of 37 households[6] averaging 4.6 persons each, there was a household average of: 2 bicycles, 0.9 televisions (a few color), 1 sewing machine, 2.3 watches, 1.6 radios, 0.4 tape recorders, 0.16 washing machines, 2.6 wardrobes, 1.3 sofas (or a pair of soft cushioned chairs), 1.4 electric fans (standing or table fans) and 0.56 desks. The village had four functioning water faucets on the street (one non-functioning) and one in the party-village council office compound. Most households also had a pump in the courtyard accessing water even more saline than the already-quite-salty tap water. They used it to wash clothes, water plants and flowers, and feed animals raised in the compound.

Fengjia Village has no day care center, but it does have a kindergarten (*yuhongban*) with 20 to 30 children and one supervisor. The primary school for grades one to five (beginning at age seven *sui*) has over one hundred students, with near universal enrollment. About half of these test into junior middle school (grades six to eight) in Sunzhen, to which they commute. (The county average is 75 percent). Only a handful continue on to senior middle school in Zouping, where they must board.

The village has a one-room health center staffed by two men: one trained in a medical senior middle school prior to the Cultural Revolution, the other an older doctor of traditional Chinese medicine. Responsibility for family planning and the distribution of contraceptives is located in Sunzhen. Three childless widows receive the "five guarantees": the village provides 400 *jin* of wheat, 200 rmb cash, 10 kg. of oil and charcoal and cooking fuel yearly, as well as medical care and funeral expenses.

Fengjia Village does not have a regular market. It lies in a territory of three standard market towns (Wangwu, Huili and Jiuhu) and one intermediate market town, Sunzhen. It is closest to Sunzhen. Table 1 depicts the 10-day schedule for Fengjia Village's nearby and farther markets.[7]

FENGJIA VILLAGE'S ECONOMY

On the wall in the village office building and in the central room of Feng Yongxi's house, where ancestral tablets would

Table 1
FENGJIA VILLAGE'S MARKETING SYSTEM

Day in Cycle	Nearby Market	Farther Market(s)
1	Wangwu	Handian; Zouping small
2	no market	Qingcheng
3	Huili	Weiqiao; Zouping large
4	Sunzhen	
5	Jiuhu	
6	Wangwu	Handian; Zouping small
7	no market	Qingcheng
8	Huili	Weiqiao; Zouping large
9	Sunzhen	
10	Jiuhu	

be, hang identical framed calligraphy in nonsimplified characters in the hand of a visiting dignitary that proclaim "the collective (road) to prosperity" (*jiti zhifu*). The continuing strength of the collective over economic activity, in the face of the dismantling of the commune structure and general retrenchment of party-state control over economic activity, was a hallmark of Fengjia Village in 1986.

Fengjia Village had an almost exclusively agricultural economy until very recently. In 1982, agriculture accounted for 90 percent of gross village income. In 1985, it had dropped to 47 percent (64 percent of net income). This has been the result of a cautious approach by the leadership to introduce industry step-by-step and of the fact that no peasants have come forward to establish their own industrial enterprises. Total village income for 1984 was 1,468,952 rmb; for 1985, it was 1,896,144 rmb (See Table 2).

Each plot of land has a fixed crop rotation (*chakou*) where one crop is sown as the other is harvested. The *chakou* are corn and wheat (1,200 *mu*), cotton (1,014 *mu*) peanuts (160 *mu*), and other crops such as hemp, millet and soybeans. Average yields, prices and costs per *mu* are, for 1985-1986 shown in Table 3.

Table 2
1985 INCOME FOR FENGJIA VILLAGE
(Figures in rmb)

Source	Gross Income	Expenses	Net Income
Total	1,896,144	822,508	1,073,636
Agriculture	883,525	200,800	682,725
Trees	44,500	1,560	42,940
Livestock	6,000	500	5,500
Industry	757,119	334,900	422,219
Household	205,000	30,000	175,000
Administrative		1,083	
Accumulation Fund		253,665	

Table 3
1985-1986 GRAIN PRODUCTION

Grain	Yield/*mu*	Price	Costs
	(*jin*)	(*yuan/jin*)	(*yuan/mu*)
Wheat	900	0.25 - 0.29	70
Corn	800	0.18 - 0.20	70
Cotton	150 - 160*	2.00**	80
Spring Peanuts	600 - 800	0.50***	60
Summer Peanuts	350 - 500	0.50***	60

* with seeds removed (*pimian*)
** breaks down to 1.80 rmb for cotton and 0.20 for seeds
*** in the shell

Per capita income figures are rather complicated (See Table 4). Prior to 1985, the figure represents the value of distributed (*fenpei*) ration grain (*kouliang*) and cash, excluding household sidelines, which were minor in any event. (Fengjia Village had no private plots after 1965 because the value of distributed ration grain, at more than 400 *jin*, was high for the region, and vegetables were grown collectively). The cash portion of incomes in the table

Table 4
PER CAPITA INCOME

Year	Income (rmb)
1978	153
1979	270
1980	500
1981	600
1982	630
1983	630
1984	632*
1985	954

* 520 rmb of this was cash; if sidelines are included, it would have come to 780 rmb.

increased yearly. Prior to 1978, the average work point ranged from 0.07 to 0.10 rmb. Starting in 1985, this figure is all cash; it does not include a *kouliang* value. The per capita wheat consumption in 1986 was around 500 *jin*.

There are officially 435 "labor powers" (*laoli*) in the village: old people and children, ages 15 to 18, count as half power; 266 (62 percent) engage primarily in agriculture—of these, 221 work mainly in grain production, and 45 raise animals or vegetables; 138 (32 percent) work primarily in industry and sidelines—of these, 88 focus on industry. Of the remaining 31, 11 work on trees and orchards, and 20 in administration, including six teachers and two health workers.

COLLECTIVE CONTROL

As China implements the ARS, the balance between collective and private (household, individual, or group) activities has become fluid. A cadre responsible for agriculture in Zouping broke down that county's collective-private balance as follows: As of 1986, about 25 percent of the villages had achieved "good" coordination resulting in rapid development and per capita income in the 800-1000 rmb range. He put Fengjia Village in this category. In another 50 percent, the collective was limited to administrative tasks (*guanli*) such as irrigation, maintaining seed quality, pest control, science and technology, provision of information, and production and sale of raw materials, but it ran (*jingying*) few of its own enterprises. Family enterprises were done well. Per capita incomes in these villages ran from 500 to 800 rmb.

The remaining 25 percent had lost coordination (*shitong*) and per capita incomes were often less than 300 rmb, only slightly above what is necessary for adequate food, shelter and clothing.

FENGJIA VILLAGE'S AGRICULTURE

According to the same cadre, grain and cotton production, prior to 1984, was based on orders (*zhiling*) but since then is based on guidance (*zhidao*). The state sends plans to Shandong Province which sends them to the prefecture and county, specifying the amount of grain to produce but not the area of cultivation. In other words, production quotas still exist. Around the Chinese New Year, the county calls together cadres from the county, *xiang* and village to meet on the plan. Over the course of three or four days, they are familiarized (*tigao renshi*, literally, raise their consciousness) with the national situation and plans. There is then a one day confab at the *xiang* attended by the village party secretary, village chief (*cunzhang*), and possibly some household representatives. They examine recent production figures, put forth a plan to allocate responsibilities, and then have discussion. When assigned quotas are agreed upon, a meeting is held in the village to relay details.

Next, in Fengjia Village, a committee of three people, one from each sub-village group, selected by the masses, writes numbers on pieces of paper up to the total number of households in the village. These are crumpled up and the head of each household picks up a piece. This indicates his priority number. The three number-writers draw last, in front of the others, to ensure fairness. Prior to this, a committee of six people from each group, plus the village council, met to rank the plots of village land as good or less good, mostly based on quality and productivity. Household heads then select two plots of land, one good, one less good. As a result, each household's plots of land are not contiguous.

Since 1985, Fengjia Village adopted a policy of assigning 2.3 *mu* of land to every villager with few exceptions. This assignment will not change for three years (*sheng bu tian, si bu jian*, literally, no additions for births or subtractions for deaths). Each *hu* can mark its land, but is not allowed to set up any sort of physical boundary that would get in the way of farm machines. The cadres insist this rule is not designed to prevent peasants from developing a sense of private land ownership, although they admit that the concept has not died out entirely. Fengjia Village has had no one wanting to switch or sub-lease land (*zhuanrang*), although it is permitted in theory.

In Fengjia Village, each individual receives land to work and each household signs a contract yearly. With no apparent irony, several cadres remarked that although it is voluntary, everyone must sign a contract. This includes cadres and those who work in village enterprises. The amount is arrived at by dividing the quota (*renwu*) assigned by the *xiang* by the number of villagers. For 1986, the quota of 194,000 *jin* of grain worked out to 188 *jin* per person. The state encourages them to fulfill the contract in winter wheat (harvested in May) and specifies this in the contract. If unable to do so, the farmers must report to their superiors. Secretary Feng said that most villagers fulfill their contracts with half grain, half something else, usually cotton.

Villagers in Fengjia sign a Zouping County Contract to purchase grain and oil. It specifies amounts, but not prices. The *"jia"* party is the grain agency at Sunzhen; *"ji"* is the household head. The "supervisor" is the village council.

Contractors are responsible for procuring seeds, fertilizer, insecticides and other chemicals. Herbicides are not used in Fengjia. Fees for "five guarantee" recipients, water projects, infrastructure, teachers and cadre wages come from profits from collective enterprises, not additional payments by households.

Agricultural tax is based on the productivity of the land and the number of *mu*. In Zouping, it averages four to six rmb per *mu* a year; in Fengjia Village, it is under four, a rate set long ago but subject to change. In 1986, tax revenues totalled 8,320.60 rmb. Households make one or two payments. They also pay 15 rmb a person to the *xiang*, a rate set by the *xiang* which uses these monies to cover its expenses. Farmers pay six rmb after selling their wheat, the remaining nine in the fall.

To this point, the strength of the collective has been demonstrated by its ability to compel nearly all residents to assume responsibility for cultivating grain land and all households to sign grain contracts. In Fengjia, the collective goes further to organize production and ensure adherence to plans and contracts.

The village monopolizes farm machine ownership and at the proper time sends them to plow, seed and harvest the land for a fee. Plowing for all four main crops, seeding for wheat and cotton, harvesting of wheat, soybeans and millet is all mechanized. The flatness of the terrain facilitates this relatively high level of mechanized agriculture. Fengjia Village has four collective service groups which both facilitate farming and tie farmers to the collective:

1) The supply and marketing group of five people purchases seeds, fertilizer, agricultural chemicals, diesel oil, lubricant oil and coal which it sells to the peasants at slightly above cost. The price is usually below that of the supply and marketing coop in Sunzhen to which some Fengjia peasants belong. As there is no need to transport materials, the

peasants turn to the local group. It also sells some local items, like peanut oil and apples, elsewhere in order to purchase goods in short supply, e.g., high grade fertilizer for Fengjia. Secretary Feng's wide connections facilitate such deals.

2) The irrigation group of 56 workers brings water from the four man-made rivers surrounding the village, as well as local well water, to irrigate the fields at set times. It charges two rmb per *mu* of irrigated land for river water and three to four for well water.

3) The livestock group of 18 workers manages Fengjia Village's 155 head of animals—mules, horses, donkeys and oxen. The animals are kept in three corrals. Ten peasants are responsible for feeding and caring for one animal which they also can use for a variety of chores, including transport. They also divide up the feces. There are only two privately-owned draft animals in the village. (In 1987, this group was disbanded and the livestock sold off to village households.)

4) The science and technology group of five members engages in research, education, technology transfer and dispersal of market information.

The workers in these groups are paid by the collective on the basis of discussion and evaluation (*pingyi*) by their fellow peasants and fees charged. In addition they are responsible for their own 2.3 *mu* of land. Fengjia Village also has a transportation team with three trucks primarily serving the village, but available for outside work.

At all stages of agricultural production then, the peasants of Fengjia Village are tied closely to the collective. When it comes time to dispose of their noncontracted crops, here, too, they find themselves drawn to the collective.

FENGJIA VILLAGE'S ENTERPRISES

With the exception of the sesame twist and cruller maker and the slaughterhouse, every other enterprise in Fengjia Village is owned collectively by the village. They were established with the purpose of utilizing local agricultural output, thereby ensuring peasants an outlet for their surplus production, keeping the operation and funding within the village, and integrating its economy. The village pays the taxes to the state. Management of the enterprises assumes several forms.

Subcontracting (chengbao)

In early December 1985 the village council sent out a notice that there would be a bidding meeting for several of the enterprises it had established in years past. It specified a minimum bid and the terms for each enterprise. Interested parties—usually groups—attended and shouted out bids. If the village's minimum asking price was too high, it was lowered. The winning groups had two weeks to work out arrangements among themselves and to sign the contract with the village. Fengjia set a three-year maximum for these contracts. In the case of tied bids, it selected the one for the longer term. Many of the enterprises had been subcontracted the previous year for just one year as a trial operation. These enterprises are known officially as "unified enterprises" (*lianheti*) as they involve more than one household, but, unlike comparable enterprises elsewhere, they are not private.

Popsicle plant: Fengjia Village's first factory, it was established in 1982 with a fixed capital of 120,000 rmb. Three people contracted it in 1985, agreeing to pay 80 percent of the profits to the village. The village provides the site, plant, raw materials and electricity. The plant has 18 employees, but only operates from May to August. Wages are four rmb for a 10-hour day, with workers each manufacturing one thousand lollies an hour. Gross production value in 1985 was 50,000 rmb, with 30,000 in profits.

Flour mill: Established in 1982, 10 people contracted it in 1986 for one year, agreeing to pay 28,000 rmb to the village which supplies the factory, equipment, a tractor, and an initial inventory of wheat. They comprise the entire workforce as well. The previous year another 10 people had contracted for 38,000 rmb but lost money so the asking price was lowered. Peasants bring their own wheat, pay to have it milled, and then store it there or take it home. Eighty percent of its income comes from non-Fengjia peasants. Of the five other mills in Sunzhen, this is the only non-private one and is the largest with a capacity of 1,000 *jin* of wheat per hour. The contractors estimated 1986's gross at 35,000 rmb, net at 33,000. After paying 28,000 to the village, they would divide 5,000 rmb among themselves based on the hours worked.

General store: Five young men subcontracted the store on December 1, 1985 for three years. They are the only employees. They agreed to pay 14,000 rmb a year. The village provides 60,000 rmb inventory, plus the building and counter. The previous year, one household had contracted to run it for 12,000 rmb but this group outbid it in 1985. Four of the five men had bid and lost in 1984. They get stock from a variety of wholesalers in Zouping and could go elsewhere if motivated. They estimated 1986 gross at 20,000 rmb leaving them 5,500 rmb to divide by hours worked. They also maintain a booth at Sunzhen.

Mechanized mill: Two young men contracted in December 1985, for one year, agreeing to pay 3,450 rmb. They are the only workers. The year before, contractors paid 3,000 rmb. These two men outbid seven or eight competitors. The village supplies equipment, but they pay for the electricity to run the machines that process wheat, millet and corn for flour, breakfast gruel and fodder respectively. They estimated 1986 gross to be 6,000 rmb with 2,000 in profit to be divided equally. Approximately 70 percent of their customers come from other villages.

Steamed bun (*mantou*) maker: Five people contracted for one year at a cost of 4,800 rmb, and are the sole employees. The year before, seven people had done it for 4,100. The major figure (who had been a team leader), had run the slaughterhouse with two others the year prior. These two wanted to switch their line of work in hopes of earning more money. In 1985, the slaughterhouse brought them 1,500 rmb apiece; they estimated clearing 1,000 rmb for 1986. The pay is based on hours worked. They sell about 20 to 30 percent of their production (daily output is 710 buns) in Fengjia and go by bicycle to other villages to sell the rest; they do not go to markets. As a rule, they do not usually sell the *mantou* for cash but in exchange for wheat: one *jin* of *mantou* for 1.2 *jin* of wheat. Cash price is 0.30 rmb per *jin*.

Fodder processing plant: No one contracted this plant in 1984-1985 but two people took it on, without a bid, for 1,000 rmb for 1986.

There are two subcontracted agricultural ventures. Apple orchard: Seven people took a three-year contract for the 43-*mu* apple orchard in 1985. They will pay 14,500 rmb plus 10,000 apples. They estimated gross income of 23,000 rmb, a net of 6,900, with division based on hours worked. Because this is agricultural work, there is no tax. They do not have a marketing contract with the state and intend to sell at markets and to peddlers who come to them. Vegetables: The village has three fields of 12 *mu* each, but of different quality. Three households contracted one field for 200 rmb; four pay 160 for a second, and three households took a third for 170 rmb. The village only supplies the well water. The growers sell vegetables in the village and at periodic markets.

Joint Operation

The cornstarch plant, established in 1984, splits profits with the village 50-50. It has 300,000 rmb in fixed capital. The manager is Fengjia Village's deputy party secretary. The plant buys about four million *jin* a year of corn from Fengjia farmers at the market price. It sells cornstarch through the supply and marketing group, mostly to factories. The gross for 1985 was 480,000 rmb; profit was 60,000. They estimated a better year for 1986: 720,000 gross and 133,000 profit. There are 36 employees, one of whom comes from Sunzhen, the only outsider involved in a Fengjia enterprise. He is the purchasing agent. The wages depend on profits and for 1985, they averaged 1,750 rmb.

Village Enterprise

The peanut oil crushing plant was set up only in January 1986, and entered production at the end of May. The village invested its own funds, not taking out any loans. It has 80,000 rmb of equipment (not including the building) and 22 employees. Most of the raw materials are locally grown peanuts. Some are pressed and the oil returned to the peasants for a fee. They also market oil in Jinan and peanut cakes for fodder via the village marketing group to state companies and on to Japan. The manager and his deputy were former team leaders. Plans call for this plant to be subcontracted once a value can be assessed. There had been a cotton oil press on this site but its equipment was sold to a private entrepreneur in another village.

Self-operation (ziying)

The slaughterhouse is run by three men at a site provided by the village for 1,000 rmb yearly rent. They signed on for one year as a trial. Because the village has no financial involvement, this is not considered a subcontract but self-management. The three butchers had no prior experience and have learned the trade as they go along. They invested 200 rmb apiece on knives and hooks. They used their savings for this purchase. They buy pigs in Fengjia and elsewhere daily (0.60 to 0.70 rmb a *jin*) and bring them to Fengjia Village for slaughter. They have no refrigeration. In summer they kill four in five days; in spring and winter, two per day. Unless the weather is bad (in which case they set up a booth outside the general store in the village square), they take the meat to periodic markets for sale. They arrive with pigs in two pieces which market inspectors stamp for a one *yuan* fee. They also pay a three rmb monthly market administrative fee and a tax of seven rmb per 100 *jin*. Pork sells for 0.78 per *jin* in rural areas, 1.20 rmb in Jinan. The butchers divide the profits daily and estimated they could clear 1,200 apiece for 1986.

*Specialized Households (*Zhuanyehu *or* Getihu*)*

There is one private enterprise in Fengjia Village. A married couple put up 400 or 500 *yuan* of their savings in 1983-1984 to make sesame twists (*zamahua*) and crullers (*youtiao*). Their customers come to them. They grossed 3,000 in 1985, netting 2,000 rmb. They estimate 1986 figures at 5,500 and 4,000 respectively. They buy wheat on the market and grind it in the local mill. They do not have agricultural obligations. They pay a three percent sales tax and a 1.50 rmb industry-commerce administrative fee.

Sidelines

While a few dozen villagers have subcontracted enterprises and several dozen more work at least part-time in these enterprises or the village's service groups, nearly all of the households appear to engage in money-earning sideline activities. A survey of 37 households showed an average of 4.9 pigs and 11.8 chickens per household. The pigs were almost all for sale, while the chickens were generally to provide eggs and meat for home consumption. The cost of raising a pig is about 30 *yuan*, the sale price for a 150 to 200 *jin* porker is 100 *yuan*, for a profit of 70. The cost is relatively low because of Fengjia's corn surplus. Elsewhere, costs may run from 50 to 60 rmb.

Three of the households were still raising Angora rabbits from a West German breed. This is down from a few years ago. When world rabbit fur prices were high (up to 120 rmb per *jin*), a number of peasants began to raise them for eventual export, but when the market collapsed (it was 40 rmb per *jin* in 1986), many ended up as dinner or on the trash heap. This illustrates the fact that increasingly, relatively remote areas of China are being integrated into the world economy, with the attendant risks. One woman earned a little money sewing hems, another making clothes.

STILL ON THE COLLECTIVE ROAD

Several things stand out from the preceding discussion of Fengjia Village's economic activity:

First, agriculture is the main source of net income as well as employment. With the exception of a handful of people, everyone in Fengjia is obliged to raise grain and sell a portion of it to the state. The village organizes production and takes responsibility for many of the chores. This reduces labor intensity and some of the more onerous tasks, freeing people for other pursuits, but precludes individual decisions about what to grow. It is a household responsibility system with the collective firmly "guiding" every household's economic decisions.

Parenthetically, nearly all subcontractors and most of the wage workers were men, causing a very obvious feminization of agriculture. Men did help out in the fields after their other jobs, but the primary burden fell on women. Women were also responsible for most household chores. These were also age stratified, so families with elderly members had an age and gender-based division of labor. The seniors handled housework, cooking and child care, in particular, enabling the wife to concentrate on farming. In any event, all family members contributed labor.

Second, nearly every household does derive additional income on the side, mostly raising pigs. Those households active on the side expected significantly higher incomes in 1986 when compared with 1985. But the pigs are all sold to the slaughterhouse or state agencies, that is, not on the free market; while there is risk involved, it is less than engaging in more purely market-related endeavors. It cannot compare with what one sees in the southern part of the country.

Third, while many people in groups subcontract village enterprises, they do not commit any of their own capital to the venture and they cannot buy out their company. They cannot develop a sense of ownership of the means of production. While they stand to lose money, it is against future income, rather than a risk of one's own accumulated savings or borrowed funds. They pay the village at the end of the year. The enterprises come with inventory, so no cash outlay is required prior to the onset of business. One could conceivably operate exclusively off the cash flow. At least into 1986, the income from such ventures was not a major part of gross family income, so that they could not qualify as "specialized households." The difference in incomes between households with or without such subcontracts was no greater than what would occur at random in the general population, according to a T-test performed on data from the survey.

Fourth, the leadership has created a rather self-sufficient, highly-integrated economy requiring limited interchange with the larger economy; most of this is filtered through collective enterprises and the village's own supply and marketing group. Peasants sold most of their surplus wheat or corn to the village processors and contracted cotton to the state. Fengjia Village's "market socialism" is very much a state- (as represented by the collective) created and dominated market. From interviews, it seemed that people go only rarely to the nearby Sunzhen market, to say nothing of further periodic markets. When they go, it is more often to visit relatives (many Fengjia brides are from Sunzhen) than actually to shop or sell. A number of itinerant vendors, repairmen, and craftsmen regularly stopped in Fengjia on their circuit to serve their clientele, obviating the need to venture forth.

Fifth, although the savings rate is high—with an average of 2,600 rmb in interest-bearing accounts at the local credit coop per household in the survey—almost all of it is earmarked for consumption—a wedding, color TV, washing machine—and not for investment in enterprises or agriculture. A few households were saving for their children's education. Some referred to themselves as "10,000 *yuan* households" if one added up the value of their assets. Fengjia Village has no one with a bank loan to repay. There is a lot of interest-free borrowing between family members and friends.

CONCLUSION

Fengjia Village differs greatly from the type of village described at the beginning of this essay. Without making normative judgments as to which type is better or normal, it is fair to ask why Fengjia is still so highly collectivized in a tide of return to the household economy.[8]

First is strong, effective leadership. Feng Yongxi has been secretary for over 25 years, through good times and bad, and is still clearly the man in charge (*yibashou*). He is a delegate to the Shandong Provincial Party Congress as well as to People's Congresses at the national, provincial and county levels. He has wide connections and strong backing from higher-ups. He has frequently consulted with county-level cadres over the village's development and they have assisted at every step. He believes strongly in collective agriculture and is being cautious and hesitant in decollectivizing entirely. He claims that he consulted with the villagers about the best system for Fengjia Village and that they support his go-slow approach. The village's flat terrain facilitates mechanization and, hence, collective agriculture; but that is not sufficient to explain why Fengjia did not just decollectivize to go along with the historical current. Feng Yongxi is central.

Second, collective dominance and Feng's cautious approach have had a big pay-off. Fengjia Village is hardly static. The economic structure has been changing, rapidly, and individuals are taking on new responsibilities, all with very little apparent disruption but with noticeable improvements in their material lives. While the village unit has established ties with the larger commodity economy, it also buffers most of the peasants in a cocoon away from market forces. The village supplies most of their needs and buys their output. This begs the question of whether they might not do even better, materially at least, if the collective was weakened further. This seems to be the direction, so an answer will be forthcoming.

Third, the Fengjia farmers did not seem "entrepreneurial," especially as compared with the south. Questions on the survey designed to get at this, such as amount of household savings, plans for these savings, frequency of trips to the market, use of free time not organized by the collective as in the past, elicited nothing that revealed an entrepreneurial urge. Possibly, once they have made their major purchases of consumer goods, they will consider investments in enterprises. Lacking overseas Chinese connections to supply these new treasures, the consumer wind has been self-funded and had to await financial solvency. In another contrast with southern China, Shandong also has little in the way of Overseas Chinese investment which could assist entrepreneurship. Shandong's overseas emigres went to Korea and they are still not in a position to bring funds back. Some of Taiwan's most successful first generation entrepreneurs were Shandongese, a fact that people still in Shandong found curious.

Fengjia's farmers appear non-entrepreneurial, even for Shandong. I interviewed several successful *getihu* from elsewhere in Zouping, and some nearby villages are retrenching from agriculture and specializing in toto in other endeavors, such as timber, flowers, furniture and food processing. A recent *Beijing Review*[9] article describes burgeoning private ventures elsewhere in the province. Fengjia Village's farmers have time on their hands, but apparently are not motivated to direct it towards risky start-up ventures, at least not as of 1987. They work hard and seek avenues to raise family incomes, including possibly risky subcontracting, but do not manifest a strong entrepreneurial urge.

This paper has described the situation in one northern China village as presented to me during a limited and highly structured visit. Other than short forays to nearby villages, I had little opportunity to gather detailed comparative data. I do not believe Fengjia's experience is generalizable. The party secretary's firm control, competence, and vision, the backing and support of higher-ups and the county's willingness to allow a foreigner to reside there, all create a sense that it is definitely a model village, something others might aspire to but few successfully copy. Chinese cadres repeated one phrase in 1986: "stimulate individual initiative" (*diaodong geren di jijixing*). Clearly, for Fengjia Village, this does not entail going one's own way, but rather, putting forth effort in activities structured by others that have a guaranteed return.

NOTES

1. Support for this research came from the Committee on Scholarly Communication with the People's Republic of China (CSCPRC). I wish to express my gratitude to the CSCPRC, as well as to the Chinese Academy of Social Sciences and the Shandong Academy of Social Sciences for their assistance. Jing Xiangkun provided invaluable support. Li Shanfeng assisted with data collection in the field, Shieh Gwo-shyong assisted at Berkeley. Justin Yifu Lin and David Zweig offered helpful criticisms in Manila.
2. See, for example, Jean C. Oi, "Commercializing China's Rural Cadres," *Problems of Communism*. XXXV(5) (September-October 1986), pp. 1-15; William L. Parish, ed., *Chinese Rural Development* (Armonk: M.E. Sharpe, 1985); Elizabeth Perry and Christine Wong, ed., *The Political Economy of Reform in Post-Mao China* (Cambridge, Masschusetts: Harvard University Press, 1985).
3. There are actually 41 village administrative councils as one village has two councils.
4. See Guy S. Alitto, *The Last Confucian: Liang Shu-ming and the Chinese Dilemma of Modernity* (Berkeley: University of California Press, 1978) and Philip C.C. Huang, *The Peasant Economy and Social Change in North China* (Stanford: Stanford University Press, 1985).
5. Some comparisons: Shandong's average per capita income is 400-500 rmb; Baijia Village's is 820; Xiaotian Village's is 4,000 (unclear if this is gross or net); Shihe Village's is 380; Lipo Village's is 504 rmb. My research assistant said that the per capita income in the Yantai region where he is from is much higher than Zouping.
6. The households were selected for me by the village leadership. I had requested a "representative" sample of the village. They chose some families whose main economic activity was farming, as well as some who worked in village enterprises or who had subcontracted them. While not "random," it did appear to be a good cross section of village residents. I interviewed 24 in their homes and another 13 in an office.
7. The language comes from G. William Skinner, "Marketing and Social Structure in Rural China," *Journal of Asian Studies*, Part I, XXIV(I) (November 1964), pp. 1-43.

8. See, for example, Victor Nee's article about a village in Fujian: "The Peasant Household Economy and Decollectivization in China," *Journal of Asian and African Studies*, XXI(3-4) (July and October 1986), pp. 185-203, and Yuan Enzhen, ed., *Wenzhou Moshi yu Fuyu zhi Dao* (The Wenzhou Model of Economy and the Road to Affluence) (Shanghai: Shanghai Academy of Social Sciences Press, 1987).
9. Duan Liancheng, "Open Policy and Cultural Ferment" (III), *Beijing Review* 46 (November 16-22, 1987), pp. 27-32, especially pp. 29-30. He discusses Mouping County near the burgeoning port city of Yantai.

THREE

THE CHINESE VILLAGE, INC.

Jean C. Oi, Harvard University

Unquestionably there has been a great transformation in the Chinese countryside since 1978. Communes have been disbanded. Collective property has been divided and contracted to private individuals under the responsibility system. Peasants are free to leave agriculture to become entrepreneurs or workers in commerce, the services, or in industry.

Will this transformation lead to the end of the collective and the re-distribution of income? The rise of 10,000 *yuan* households, reports of empty collective coffers, decreased collective welfare funds for village schools and clinics, the end of health insurance, and the decay of village irrigation systems and roads, would certainly suggest such a fate for the corporate village.[1] Yet many villages remain strong corporate entities. They continue to re-distribute substantial amounts of income to their members and give members a stake in remaining part of the collective.

This paradox suggests that it is not the contracting of property or the rise of the household economy that necessarily leads to the demise of the village as a corporate entity. The issue is not the reforms as such, but the sources of income in a village, and more specifically, the degree of industrialization in a village. As later sections of this paper will show, villages with highly developed industry are able to provide impressive subsidies to villagers, whether they work in rural industry or remain in farming.

Villages that rely exclusively on agriculture, particularly in the more remote areas with little or no industry or other income generating activity, are the ones most likely to be weakened as a collective. Such villages are left with few (legal) income sources, since they get neither a share of the agricultural tax nor the profits from peasant sales of grain to the state—the only major assessments on farm income. Peasants have autonomy over the management of their work and direct control over the profits from their harvests. Thus villages that rely primarily on agriculture have experienced a decline of the collective both in fiscal and organizational terms.

In contrast, villages that have non-agricultural enterprises still control them and redistribute their income. This income from industry, which in some villages may be many times greater than that from agriculture, provides a collective cohesion that some have claimed will disappear with the household contracting system and private enterprise. Consequently, to assess the impact of the reforms, one must look not simply at what has happened in the organization of agriculture but at what has happened to the industrial sector of the rural economy as well.[2]

Evidence suggests that industry, not agriculture, will become the primary source of income in large areas of China's countryside. Already in some suburban villages, industry is the primary source of both family and collective income. In these villages, income from agriculture amounts to less than half of the total. In some it is less than a third, and in others less than 10 percent.[3] For example, in one village in Shenyang, agriculture contributed 7.3 percent to the total income, while industry contributed 63.6 percent.[4]

Christine Wong has shown that "by 1986 the gross value of output to rural enterprises exceeded the gross value of agricultural output for the first time," with the number of rural enterprises growing rapidly, and quadrupling in 1984.[5]

Moreover, as work in agriculture becomes less lucrative than in industry and services, more peasants want to leave farming.[6] According to official statistics, in 1984 100 million peasants were in non-farm activities, many in rural enterprises.[7] By the end of 1985, township enterprises had absorbed some 60 million surplus laborers.[8] Wong found that the numbers employed in these enterprises rose from 32.4 million in 1983 to 69.8 million in 1985.[9] In highly developed areas such as Wuxi, Jiangsu Province, 40 percent of the rural labor power is engaged in industrial production, with almost every family in the area having at least one member working in a factory.[10] Even in those areas that are not highly industrialized, an effort is made to have at least one member of each household work in the existing industrial enterprises in an attempt to even out household incomes.[11]

The loss of control by village governments over the day to day management of agricultural work and the rising autonomy of households that comes with the responsibility system in land affects only a small part of the village economy. As the following sections will show, the decentralization that has taken place in agriculture has not taken place to a similar degree in the management of rural industry. The responsibility system has been instituted and factories have been contracted out, but villages still control their enterprises through a web of rules and regulations that govern the division of enterprise profits, access and allocation of investment opportunities, and key inputs and credit. The main contention of this paper is that the corporate nature of the village will rise with the level of industrialization. The following pages will detail why this is so.

THE VILLAGE ENTERPRISE CONTRACT SYSTEM

To understand how village government is able to maintain effective control over the financial flows of village industry, one must first understand the terms of the contract responsibility system for rural enterprises. On the surface the responsibility system is deceptively simple—the contractor pays taxes to the state and a share of the profits and perhaps a management fee or surcharge to the village or township government who owns the enterprise. Once that obligation is fulfilled, one assumes that contractors of village industry enjoy the autonomy of management and rights to profits similar to those enjoyed by peasants who contract farm land. The relationship between contractors of enterprises and the village is, however, much more complex than the bare outlines of the responsibility system would indicate.

The term "contract" is used to describe the relationship between the government and the individual entrepreneur, but the use of the term is somewhat misleading for a Western audience. A contract in this context does not necessarily denote a legally binding agreement negotiated by two parties of equal status. Village governments as owners of the enterprise are unquestionably the party with the power to determine the terms of the contract. They set the rents, determine the profit margins, and as later sections will show, have the right to intervene in the internal management of the enterprise after it has been contracted out. Moreover, when they feel that the interests of the collective are no longer served by the terms of the contract, it is not uncommon for the contract to be unilaterally nullified and new terms drawn up. Contractors in this situation are often left with no recourse, even though in theory they are protected by law. This has been the case in those enterprises that have turned out to be unexpectedly profitable.[12] Increasingly, as village governments learn from the experiences of the first few years of the reforms, they are making major changes in the way that enterprises are contracted out, both in terms of rents and in terms of to whom they are contracted. Let us first look at the way profits are divided under the contracting system for industry.

RENT AND DIVISION OF PROFITS

Contractors pay either a fixed amount of profit to the village each year regardless of income or a fixed percentage of each year's profits to the collective.[13] The first method is the one many villages used, especially at the lower levels and in the earlier stages of the reforms. Villages simply decided on a set profit target and allowed the contractor to keep any over quota profit. But as the reforms took hold and governments gained some sophistication and saw how successful some of the enterprises were under private management, they soon developed more effective ways of getting income.

In the floating rent or percentage system, all profits up to a set amount are divided between the village and the contractor according to a ratio set by the local government; for example, 7:3 or 8:2, with the village getting the larger share. After this minimum profit quota is met, a second ratio is used to divide the over quota profits between the village and the contractor (*lirun baogan chaoli fenchen*). Of the over quota profits that a contractor can retain, a set percentage is designated to go back into the factory, with a portion given to the workers. Again it is the village, specifically the industrial company (*gongye gongsi*) of the village government, that sets the ratio to determine the percentage of over quota profits each party will receive. A common ratio is 2:3:5—the village gets 20 percent, the contractor 30 percent, and the factory 50 percent.[14] In this way, the village is assured of a minimum profit and still gets a share of any unexpected profits.

The shift to the percentage system occurred in a village outside of Shenyang, for example. The village government originally used the simple system where the contractor paid a fixed amount regardless of profits, and was entitled to the over quota profits. Beginning in 1985, however, this village decided to switch to a floating percentage rent. The village set the ratio for the division of profits at either 7:3 or 8:2, putting the later ratio into effect if the income was expected to be high.[15]

The village official explained the change by saying that the fixed rate had allowed some particularly successful enterprises and entrepreneurs "to stand out too much." The three levels of interest—the state, the collective, and the individual—were not in harmony. The percentage rate would ensure more equality. The same cadre pointed out that the floating system protects the contractor if problems arise that prevent him from meeting the set target. True though that may be, the decreased profitability for peasant entrepreneurs resulted in a decrease in the percentage of enterprises contracted. Plans to extend this system to all enterprises in the village, including those owned by former production teams and presently contracted out on the simple profit quota system, further suggests that stricter control will be more, rather than less, prevalent. An official said the set profit quota system was too unregulated.[16] More to the point, perhaps, it was too unprofitable for the local governments.

BONUS PAYMENTS

In some cases the village continues to control all of a factory's profits and the contractors only receive a bonus wage. This occurs when an enterprise, as a collective, contracts the factory from the village. Usually the factory manager, and perhaps his top staff, act as the principle contractors. In this case, the factory manager and his staff receive a bonus from the village government if the factory meets the targets set by the village industrial corporation (*gongye gongsi*), instead of a set percentage of the profits. This provides the manager and other cadres, such as the accountant, with a salary about twice that of the average worker. For example, in Nanjiao district outside of Tianjin, the factory manager of a collective exhaust fan factory makes a contract with the township (*xiang*) to meet production and profit targets. He gets a bonus, including such items as a wrist watch, from the *xiang* if he exceeds the targets and his performance is considered outstanding.[17]

Collective contracting with factory manager responsibility is quite common and in some villages preferred. Some fear that if an enterprise is contracted to private individuals they will not take care of the equipment, but merely try to make as much as possible off the factory the first year and be done with it. Officials also say that most enterprises are too large for most individuals to handle. Collective contracting also allows the village the most control since it is essentially only a work responsibility system that fines factory managers if quotas are not met.

VILLAGE CONTROL OF MANAGEMENT AND INCOME

The intervention of village governments does not stop with the collection of rents or management fees.[18] In contrast to their withdrawal from day to day management in agriculture, village governments continue to limit the autonomy of those who contract rural enterprises.[19] The degree to which village governments control internal factory affairs varies from area to area, sometimes from one village to the next. For instance, some factory managers who have contracted an enterprise have no say in who is hired. In the exhaust fan factory outside of Tianjin, if workers are needed, the township (*xiang*) government appoints and apportions the jobs among the villages and work groups (former production teams). The township industrial company (*gongye gongsi*) takes care of labor allocation and administers qualifying examinations to workers when necessary. The number of workers a factory may hire is controlled by the funds allocated for wages.[20] Other villages allow individual contractors to make all hiring and firing decisions.

The more important issue, from the perspective of this paper, is the consistent use of plans and targets. Economic enterprises, both those that have been contracted to individuals under the responsibility system and those that remain under collective management, are still subject to numerous plans, targets, and quotas set by the village and township governments. Some of these plans and targets are more strictly adhered to than others. Some are "guidance plans" (*zhidaoxing jihua*), others mandatory plans (*zhilingxing jihua*). The distinction, however, is not as clear cut as one might think. Guidance plans often seem to carry as much weight as a mandatory plan.

The difference seems to boil down to the means by which a target is achieved. One official in China who tried to explain the difference between mandatory and guidance plans said that, "mandatory plans are like putting a ring through the nose of an oxen and pulling it where you want it to go. Guidance plans are like letting the oxen go where it wants to go, but not feeding it if it refuses to go where you want it to go." In other words, this difference between the stick and the carrot is the difference between using administrative and economic levers. Mandatory plans are achieved through the use of administrative fiat. Guidance plans are achieved through the use of economic incentives.

The plans for the distribution of funds and tax quotas retained in a factory fall into the category of mandatory plans. A factory manager, like his urban counterpart, has little leeway to use the factory's profits as incentives for workers. Regulations specify that 50 percent of the after taxes profits must be plowed back into the factory to expand production. Only 20 percent can be used as bonuses and benefits (*fuli*).[21] Local regulations further mandate that the remaining 30 percent of the profits go to the collective to subsidize agriculture and provide for village welfare.[22] Compared to the amount of enterprise profits that went into agriculture prior to 1984, this 30 percent rate is low.[23] But compared to national statistics which show that the proportion of net profit specifically designated for supporting agriculture has been going down, from 5.2 percent in 1985 to 4.3 percent in 1986,[24] the rate of 30 percent in these suburban villages seems high.

Regardless of whether the above quoted rates are representative of more general patterns of the amount of profits going into agriculture, the total amount that local governments extract remains high. Nationwide, in 1984 *xiang* and *cun* village governments commanded an overall 43.9 percent of after-tax profits from rural enterprises.[25] Wong captures well what many factory managers must still feel; that they are treated like "little cash registers" by their local government.[26]

Other plans, such as that for product line, are mandatory for some factories and guidance plans for others. Those factories that produce "key items" and supplies which are provided by the government at state prices must follow the state's plan of production and sale. For example, the exhaust fan factory outside of Tianjin gets 80 percent of its supplies from the state; specifically, the Tianjin planning commission, so the majority of its production is dictated by the state's plan.[27]

It is much less clear how closely a factory must adhere to its "guidance" production plan. What is clear, however, is that those factories who adhere to what is desired by the local governments have the best chance of success. Local governments can give selected factories that decide to accept the carrot rather than fight the stick crucial assistance in the areas of loans, investment opportunities, provision of raw materials, and in markets for finished goods. The result is a mix of regulation and a continuation of the "iron rice bowl" paternalism that characterized the pre-reform urban state owned sector.[28]

Village governments can, for example, affect the taxes and profits of an enterprise by helping the contractor/manager obtain special consideration for credit and investment loans, some at low or no interest. Individuals may contract to operate enterprises, but because these by definition are rentals, entrepreneurs will seldom, if ever, invest in these ventures with their own funds. All investment in collective owned enterprises are made by the collective regardless of who actually manages the operation, the collective or individual contractors.

Consequently, when factories want to expand production or buy new machinery, the contractor or manager goes first to the village or township governments, not to the bank. Depending on the size of the investment or loan needed, the village or township will either provide the funds out of its accumulation fund or, for larger amounts, arrange for a loan from the local savings cooperative or branch of the Agricultural Bank. Loans are also available from the finance and tax offices. Whether or not these loans are approved depends not on the performance of the individual entrepreneur but on the village officials, and, in the case of larger projects, on the approval of higher level officials at the township and perhaps the county.

The relationship between banks, finance and tax offices and village officials is very close. Without a nod from the basic level officials, a loan is almost impossible. The village acts in some instances as the guarantor of a loan. In other cases, an enterprise may find another to act as its guarantor, but ultimately the village officials must approve it, at least informally. Through the intervention of village or township government, an enterprise may receive a subsidy to pay the interest on a loan, or an exemption from penalty interest payments, an extension of the payback period, or even the opportunity to use tax dollars to repay a loan (*shuiqian huankuan*).

Local governments also play a major role in the fortunes of an enterprise through the supply of key production inputs and the location of markets for finished products. One of the most coveted and profitable arrangements for rural industry is to land a contract with an urban unit. This usually is an industrial unit, but it may also be a "scientific unit" such as a university or scientific lab developing new products. The urban and rural units form a "horizontal linkage" (*hengxiang lianhe*), whereby the rural partner does all or part of the production process. The urban partner is responsible for supplying the raw materials and agrees to market or purchase the output of the rural factory.[29]

Village governments can be instrumental in finding such opportunities. Outside of Tianjin, village officials arrange special tours to their factories. In Sichuan, near Chengdu, various government offices, including those at the county level, actively seek cooperative relationships for rural industries.[30] Repeatedly officials everywhere said, "*Guanxi* is important in getting business. We have built up old connections and now use them."

Village governments further help by investing in factories that produce raw materials needed by its industries. This guarantees needed supplies for its enterprises and provides profit for the village. The prices that the village charges are close to market, but the issue in China for many items is not price, but access and convenience.[31] The profit that the village makes from these sales is then used for collective welfare.

Like the corporations (bureaus) and second level companies (*erji gongsi*) that run urban industry, village and township governments function like corporate boards of directors. Cadres at these levels decide questions ranging from spending, investments and loans, to hiring. They also make provisions to assist their enterprises in acquiring credit and needed inputs. The reforms have decentralized farming, but they have done little to decentralize the management of rural industry to those who have contracted to operate these enterprises. This consequently has allowed village government continued control.[32]

THE NEW RE-DISTRIBUTIVE SOCIALISM

With the funds that villages with highly developed industry have been able to extract, they can provide impressive subsidies to villagers, whether they work in rural industry or remain in farming.[33] For example, villages have built schools, clubs, apartments and houses, movie theaters and club houses with village revenues. Some villages provide free water, electricity, and liquid fuel. A number give subsidies for education; in one case up to 1,000 *yuan* per student who tests into college, in another 600 *yuan*.[34] In the latter village, 200 *yuan* is provided for students who go to vocational schools.[35] A village outside of Shenyang provides 60 *yuan* per student, 60 *yuan* in nursery school fees per toddler, three *yuan* per person for health insurance, and 20 *yuan* per month in old age pensions for men over 60 and women over 55 years old. This village also provides each member free of charge 550 *jin* of rice as their basic grain ration, since most of the villagers are no longer engaged in agriculture.[36]

The loans and investments described in the previous section are another important way income is redistributed within the village. The village may grant poorer enterprises a loan from the collective accumulation fund or start new industries using profits from existing industry. Investments or "loans" are also made by workers in village industry. In parts of Sichuan, peasants who want to work in village enterprises pay 500 to 1,000 *yuan* to the factory. They receive 8-10 percent interest on this money, and the principle is returned if they quit. In the meantime the factory (and village) has use of these funds.[37] Wong also has described such practice, which is termed "*gongren daizi ruchang.*" She found that in Wuxi workers were required to bring in with them 2,000-3,000 *yuan*.[38]

Perhaps the most important subsidy that villages make, aside from those for general welfare, is for agriculture. The disparity between the income from industry and agriculture has forced the state to take direct action to stop the exodus from farming, particularly grain production. The policy is called "use industry to subsidize agriculture" (*yigong bunong*). Village governments are mandated to take profits from industry to directly assist agriculture.

The question is the extent to which local governments do in fact take the profits they control and invest it into agriculture. As indicated earlier, a set amount of each enterprise's profits is earmarked for this purpose, but there is considerable variation. I have already indicated that the national statistics suggest that the average amount is substantially below the percentage guidelines found in the highly industrialized rich suburban villages, outside of large cities such as Tianjin.

I suspect that the figures I obtained in interviews were less than precise and included both direct subsidies for agriculture and welfare expenditures. The disparity is reduced if one takes these two expenses together. Based on national figures, the percentage increases to 16.7 percent for 1985 but decreases to 13.2 percent for 1986.[39] As Wong has pointed out, one must also take into account the problems with national aggregate statistics, both in the way that statistics are collected, and in what they represent.

Whatever the inconsistencies and imprecision in the use of the term "funds to support agriculture," I think that it is safe to assume that the higher rates I was quoted in interviews are perhaps peculiar to the highly industrialized suburban communes, where it is clear that agriculture is on the decline and that without considerable support would diminish to the point where there would be insufficient production of foodstuffs.

The imprecision may also be due to the fact that subsidies to agriculture take different forms and it is unclear how these are accounted for. One is payment of the agricultural tax or the management fee by the village for those peasants who still farm. Other ways that "industry helps agriculture" include factories starting work late in the morning, so their workers can help with agriculture. In really rich and highly industrialized areas, villages subsidize the costs of agricultural equipment purchased by farm specialized households. Outside of Shenyang, if a specialized household wants to buy a reaper for 4,000 *yuan*, for example, the village might provide a subsidy of 3,000, and the specialized household would only have to pay 1,000 *yuan*.[40]

CONCLUSION

Overall, recent evidence suggests that the diversification and particularly the industrialization of the village economy that have followed the reforms can allow the collective to endure as a corporate entity, but one different in character. The economy of the *xiang* or the *cun* perhaps should now be thought of as that of a diversified corporation, rather than that of a simple village. Rather than having a single source of income, villages now have diversified operations, some of which are more profitable than others. Profits from one sector can be used to support another. And as in a corporation, when a division is weak, but seen as vital to the overall health of the company, profits are drawn from stronger divisions to maintain it, regardless of costs. This is the relationship between agriculture and industry. The relatively low profits in farming, and most importantly, the inability of the village government after the responsibility system to redistribute income from the agricultural sector have made the growing industrial sector the financial mainstay that keeps village finances afloat.

To what extent this is actually implemented in different villages in China is a separate question, and one which will, no doubt, yield tremendous variation in findings. The disparity in the interview data on the amounts used for the support of agriculture and actual amounts used based on the national statistics is but one indication that this policy is not being uniformly implemented. The data on which most of this paper is based are clearly not the norm. The highly industrialized suburban villages and townships are probably among the richest.[41] Obviously, not all villages are able to provide such subsidies. The point here is not to suggest that all villages will achieve this level, but that this is what can happen as a result of the development of rural industry and a conscious policy of support for agriculture. More importantly, it suggests that local governments, both at the village and township levels, have retained the ability to extract and direct investment and spending of rural industrial profits, the mainstay of village income. Whether they in fact decide to use this income to support agriculture or take these profits to expand the industrial base of the village is a separate issue. That depends on economic as well as political issues, such as the amount of pressure from the upper levels and the cost of letting agriculture die as more and more peasants move into the comparatively more lucrative industrial sector.

What I have tried to show here is that the reforms do not necessarily lead to the end of re-distributive socialism. It can and has led in some villages to a new and perhaps stronger form of collective corporatism. On the contrary, it seems that it is precisely in those areas where the reforms have been pushed to the limit, where the economy has become diversified, where agriculture has become the work of specialized households, where the contract responsibility system has flourished, and most importantly, where industry is booming and where peasants have been allowed to become 10,000 *yuan* households, that the collective is strongest and wealthiest.[42] It is in this type of environment that the collective is able to draw on new channels of revenue and income to fund a new type of re-distributive socialism.

NOTES

1. The term "village" in this paper will refer to what under the Maoist era was the collective. Consequently, this term will be used in the broader sense to encompass both the present administrative units known as the "village" (*cun*) and the township (*xiang*). I am aware of the problem of such imprecision, but the purpose of this paper is to discuss the continuing relevance and strength of the "collective," not to clearly delineate the functions of the various administrative units within the new rural bureaucracy.
2. This aspect of the reforms has received almost no attention. See, for example, the otherwise useful volume on reform in China by Christine Wong and Elizabeth Perry, eds., *The Political Economy of Reform in Post-Mao China* (Cambridge, Massachusetts and London: Harvard University, Harvard Contemporary China Series 2, The Council on East Asian Studies, 1985).
3. I spent April through August of 1986 in China doing preliminary interviews on village industry. While in Tianjin, Shenyang, Dalian, Chengdu, and Chongqing, I was able to interview various officials at the *qu, xian, xiang,* and *cun* levels, (a total of 14 sites) about changes in the planning process and the degree of government involvement in the management of the rural economy, focusing on the management and finance of rural enterprises. At the *qu* and *xian* I met representatives from the planning and economic commissions, the agriculture and forestry office, tax, price, and finance offices, commercial office, sales cooperative, agricultural bank, and the *gongshang guanli* office. At the *xiang* I interviewed *xiang* heads and vice heads, the *xiang qiye gongsi jingli*, heads of saving cooperatives, and factory managers of *xiang* run industry. At the *cun* level I interviewed *cunzhang*. Unless otherwise indicated, the material for this paper is based on these interviews. I want to stress that I am here laying out only the most *preliminary* of findings for discussion. The propositions sketched out here will be further tested and no doubt modified as I continue to research this topic.
4. China Interview 72886.
5. Christine P.W. Wong, "Interpreting Rural Industrial Growth in the Post-Mao Period," *Modern China*, Vol. 14, No. 1 (January,

1988), pp. 3-30. Wong cautions, however, that the absolute numbers should not be taken too literally. In spite of statistical problems, the real growth of the number of enterprises is nonetheless significant.
6. See Jean C. Oi, "Commercializing China's Rural Cadres," *Problems of Communism* (September-October 1986), pp. 1-15, for details on the differences in wages between those in agriculture and industry.
7. Cited in Wong, p. 3.
8. "1985, 1986 and Beyond...," *Beijing Review*, January 6, 1986, p. 4.
9. Wong, *op. cit.* The number of people working in rural enterprises varies according to the different statistical sources available. According to the statistics bureau, data in 1986 is lower than that which Wong calculated using an amalgam of sources. It lists only 43,915,300 people working in rural enterprises, out of which 21,166,500 worked in village (*cun*) owned enterprises. *Zhongguo tongji nianjian* 1987 (Beijing: Zhongguo tongji chubanshe), p. 206. The number, when available from the *1987 Agricultural Yearbook*, will probably be higher. According to Christine Wong, the figures from the state statistical bureau on this item tend to undercount the number of persons actually working in rural industry. Regardless of the precise number, the trend is clearly toward peasants moving into non-farm activity.
10. Xue Muqiao, "Rural Industry Advances Amidst Problems," *Beijing Review*, December 16, 1985, p. 18.
11. Du Rensheng, director of the Rural Development Research Center of the State Council, announced that the state plans to transfer over 50 percent of the surplus labor from agricultural production to nonagricultural departments by the end of the century. *Beijing Review*, June 24, 1985, p. 17.
12. See Oi, "Commercializing," for examples of this problem. Also see David Zweig, Kathy Hartford, James Feinerman, Deng Jianxu, "Law, Contracts, and Economic Modernization: Lessons from the Recent Chinese Rural Reforms," *Stanford Journal of International Law*, Vol. 23, No. 2, 1987.
13. The precise method varies from village to village. Tom Bernstein provides details of the process in two counties in Shandong and Anhui. See his "Local Political Authorities and Economic Reform: Observations from Two Counties in Shandong and Anhui, 1985," paper prepared for presentation at the Conference on Market Reforms in China and Eastern Europe, (La Casa de

Maria Conference Retreat, Santa Barbara, California, May 8-11, 1986). Graham Johnson describes the process for villages in the Pearl River Delta. See "The Fate of the Communal Economy: Some Contrary Evidence From the Pearl River Delta," paper presented at the Annual Meeting of the Association of Asian Studies (Chicago, March 21-23, 1986).
14. Some contracts are very comprehensive and specify the amount of profit the factory manager gets, the amount for workers, etc. Others are less specific.
15. China Interview 72985.
16. China Interview 72986.
17. China Interview 71886.
18. The management fee in parts of Sichuan, for example, is one percent of total income. This money is collected by the tax office from both village and township enterprises. The township receives the money but then gives a portion to the *qu*. China Interview 81786.
19. Christine Wong makes a similar point, *loc. cit.*
20. See *supra* note 17.
21. Based on national statistics, the actual percentage of profits used for reinvestment in 1985 was 46.3 percent. In 1986 it was 49.8 percent which is closer to the official 50 percent figure, *Tongji nianjian* 1987, p. 205.
22. This is sometimes paid to the village and sometimes to the township. In Nanjiao, outside of Tianjin, village industry pays only the township, not village, which then distributes the funds to all of its villages. Local officials said that this ensures fairer distribution. The township in this area also takes care of the five guaranteed households. See *supra* note 17.
23. Wong, *op. cit.*, pp. 18-19.
24. *Tongji nianjian, op. cit.* p. 205.
25. Wong, *op. cit.* p. 23.
26. *Ibid.*
27. See *supra* note 17.
28. Based on my interviews in urban state owned industry, much of this paternalism still exists. For a description of the continued "soft budget constraint" in state owned factories, see Andrew G. Walder, "The Informal Dimension of Enterprise Financial Reforms," in US Congress Joint Economic Committee, *China's Economy Looks Toward The Year 2000: Volume 1. The Four*

Modernizations (Washington, DC:US Government Printing Office, 1986), pp. 630-645.
29. There are exceptions to this, such as the exhaust fan factory mentioned earlier. The factory itself sold its products. This was not a problem because the fans were required in certain types of factories, so demand was high.
30. China Interview 82386.
31. See Jean C. Oi, "Peasant Households Between Plan and Market: Cadre Control Over Agricultural Inputs," *Modern China*, Vol. 12 No. 2 (April 1986), pp. 230-251, for some of the problems in this regard. That article deals with agricultural inputs, but the problems faced by rural industry for materials such as wood and cement are similar.
32. Kathleen Hartford points to the need for cadres to increase their control over finances if they are to keep the collective alive and their power strong. "Socialist Agriculture is Dead; Long Live Socialist Agriculture Organizational Transformation in Rural China," in Wong and Perry, eds., *op. cit.*, p. 60. The rules for rural industry seem to be key to this strategy.
33. Graham Johnson found that villages in the Pearl River Delta also provided substantial funds for collective welfare. See his "The Fate of the Communal Economy," *op. cit.*
34. See *supra* note 4; China Interview 81786.
35. China Interview 81786.
36. See *supra* note 4.
37. See *supra* note 35.
38. Wong, *op. cit.*, p. 25.
39. *Tongji nianjian, op. cit.* p. 205. The decrease is also apparent in absolute as well as percentage terms.
40. See *supra* note 4.
41. For example, the village in Shenyang which provided the most impressive subsidies was particularly rich. It had 26 village level enterprises. In 1985, the total production value for the village was 4,305 *wan yuan*, which was 4 times its value in 1978. Breaking this down to the contribution of different sectors of the village economy, it becomes clear how important village industry is:
Agricultural production = 7.7 percent
Animal husbandry = 8.8 percent
Transport work = 8.43 percent
Commercial enterprises = 4.6 percent
Industrial enterprises = 63.6 percent

42. I have only dealt with the collective's ability to extract and re-distribute funds from collective enterprises here, but there are also numerous ways in which the village is able to extract money from individual entrepreneurs. These methods, however, are less open and rely on more informal, subtle, and sometimes not so subtle methods. I have described some of these methods in my discussion of cadre envy (*yanhong*). See Jean C. Oi, *op. cit.* For an interesting account of the social leveling process from an anthropologist's perspective, see Ann Anagnost, "Prosperity and Counter-prosperity" (October 1986), unpublished paper.

FOUR

THE CHINESE ECONOMY IN THE NEW ERA
CONTINUITY AND CHANGE

Robert F. Dernberger, University of Michigan

The Third Plenum of the 11th Central Committee of the Chinese Communist Party, at the end of 1978, is accepted as a major turning point in China's modern history. Tsou Tang has gone so far as to argue that it was the most important political event since the May 4th Movement in 1919, while the Chinese Communist Party cites it as the most important political event since the founding of the People's Republic of China in 1949.[1] Whatever the judgment of history, that meeting certainly led to the emergence of a new political alignment in the post-Mao period. The Third Plenum also began the open challenge and eventual rejection of the Maoist economic principles and introduced the Eight Character Program of experimental economic reforms that has eventually led to significant changes in China's economic system, development strategy, economic policies, and even values and behavior. Thus, the Third Plenum marks a watershed in the history of China's economic modernization.

The detailed history of the Chinese economic reforms that have followed the Third Plenum will not be presented here; that story has been told elsewhere by the author and many others.[2] Moreover, the full story must wait to be told as the process of economic reform, i.e., institutional, strategy, policy and behavioral changes, is still unfolding and there is considerable disagreement over when that process will come to an end and with what result. Rather, our purpose in this paper is to try to obtain an assessment of the impact the reforms have had thus far, i.e., as of the present point in time. We are not trying to judge whether the reforms have been a success or failure. The Chinese are trying various economic reforms without a particular model to guide them in their search for "a socialist economic model with Chinese characteristics." The Deng/Zhao leadership has displayed considerable determination and perseverance in their reform efforts over the past 10 years; the fact that they have not yet achieved their goal and have encountered obstacles, some created by the reforms themselves, cannot be cited as proof of their failure. Our purpose in this paper is to ask, in terms of the available macro-economic indicators, what difference the economic reforms have made to the economy and its performance thus far?

While not attempting a precise assessment of how far the economic reforms have changed China's economic system, development strategy, and economic policies, this paper will demonstrate that important changes in these areas have been achieved. The mix of plan and market, decentralization and central control, state ownership and private ownership that will evolve out of China's economic reform program is likely to fall somewhere between the two extremes of a traditional Soviet-type economy and the pure market socialism model of Taylor and Langer.[3] But this still leaves a very wide spectrum of choices available to the Chinese. How far have the Chinese gone thus far in their development of a mix between old and new economic institutions, strategies, and policies? Unfortunately, even this easier question is difficult to answer in any precise or statistical manner, as our analysis in this paper will show. Each observer must rely on their own biases and intuitions in describing the glass either as half full or half empty. And anyone, of course, is free to speculate as to what is likely to happen in the future. There are those who believe that market socialism is right around the corner in China. Some believe it is already here. More cautious observers cite the dead weight of traditional institutions, bureaucrats and cadres, and patterns of behavior to argue that actual changes have been far less successful than the advocates of economic reform in both China and abroad would have us believe.

The Hungarian economic reforms were adopted as a complete package in 1968, a decade before the introduction of China's economic reform program. In addition, the Hungarian reforms included the abolition of many features of the traditional Soviet-type economy that the Chinese still retain. Nonetheless, Hungarian economist Janos Kornai argues that in reality the Hungarian economy still falls well short of the reforms objective of a market socialist economy.[4] He also questions whether economic reform actually can be achieved in a traditional socialist economy if the bureaucracy is unwilling to observe a voluntary restraint from its interference in the economy. Kornai claims, and we agree, that this question "cannot be answered by speculation, only by historical experience,"[5] and he advises us to wait and see what may be revealed by Hungarian, Yugoslav, or Chinese experience before we waste our efforts in arguments on behalf of one outcome or the other.

Thus, following Kornai's advice, we limit ourselves to a more feasible, yet still important, question concerning the impact of the economic reforms in the new "post-Mao era" in China. What is the extent to which China's economic performance over the past decade represents the continuity of past trends and to what extent does the pattern of economic performance represent a new and different era in China's economic development? To answer this question, the presentation is organized in a rather straightforward manner. We begin by examining the rate and structure of economic growth, i.e., the statistical results of growth. Then, we identify and try to measure the sources of that growth. Finally, we investigate the consequences of that growth on the life of the individual Chinese, i.e., the benefits of that growth. In the conclusion, an assessment is made of the mix of continuing trends and basic changes in the record of China's economic development over the past decade.

THE PATTERN OF GROWTH: RATES

Well endowed with natural resources and supplies of labor, the Chinese were able to achieve high growth rates after 1949; the Soviet economic system they adopted after 1949 was designed largely for achieving that objective.[6] Statistical indicators for national income in the socialist countries, of course, are not comparable with those for the non-socialist countries. Nonetheless, most countries that had Soviet-type economies achieved relatively high growth rates of physical output during the period 1965-1980 (See Table 1).

The Chinese maintained a record of growth over this period that was slightly better than that for the East European countries, especially over the last half of the 1970s, but also significantly better than the growth record of the Soviet Union, the industrialized market economies, and the other low income countries in the Third World. Quite simply, the Chinese used the Soviet-type economic system to mobilize resources for growth to a greater extent than in the other socialist countries and to a significantly

Table 1
AVERAGE ANNUAL RATES OF GROWTH OF NET MATERIAL PRODUCT[7]
(in constant prices)

	1956-1970	1970-1975	1975-1980
USSR (GNP)[8]	5.2	3.7	2.7
Bulgaria[9]	8.5	7.8	6.1
Czechoslovakia[10]	6.8	7.8	6.1
East Germany[11]	5.2	5.4	4.2
Hungary[12]	6.8	6.3	3.6
Poland[13]	6.0	10.0	1.4
Romania[14]	7.6	11.2	5.0
China[15]	8.3	5.5	6.0
Non-Soviet-type economies[16]		1965-1973	1973-1984
India		3.9	4.1
Low-income countries		3.7	4.1
Middle income countries		7.4	4.6
Industrial market economies		4.7	2.4

greater extent than was true in the market economies that relied on a more voluntary rate of savings and investment. Yet, China's experience since 1949 has taught us that high growth rates do not necessarily signal successful economic modernization. These high growth rates can be accompanied by the failure to achieve innovations and technological change, a lack of improvement in the quality and variety of output, a failure to improve factor productivity and efficiency, and retarded growth in the standard of living. In fact, excessively high growth rates may well lead to failure in the attempt to achieve modernization, i.e, Nigeria, Iran, and China itself before the end of the 1970s. Purely on the basis of the statistical results for growth in the period before the Third Plenum, however, the Chinese economy stands out as a growth economy—among the highest growth records in modern history.

Statistical analyses of the socialist economies have found

that these economies generally exhibit relatively high growth rates, but that these growth rates are rather unstable, i.e, these economies experience cycles in economic activity about the long-run trend, and the long-run trend has a diminishing slope—the growth rate declines over time.[17] These same characteristics were exhibited in China's record of growth. During the period 1952-1978, China's average annual rate of growth of national income (in comparable prices) was 7.21 percent with a standard error of 11.52, a coefficient of variation of 159.64, a moment coefficient of skewness of 1.14, a moment coefficient of kurtosis of 5.01, and a negative time trend of -0.14 percentage points a year.[18] This is the statistical description of a growth rate that follows a cyclical path with troughs centered on 1961, 1967, and 1976, around a linear time trend that slopes downward and passes through a mean of about 7.2 percent.

What has been the impact of the economic reforms on this pattern of growth? No attempt can be made here to evaluate the true or precise impact of the economic reforms upon China's economic performance, i.e, while holding other factors causing changes in that performance constant or removing the impact of these other factors. The impressions of the Chinese policy-makers are based upon their intuitive reading of the macro-economic results, but what they presently accept as the "impact of the reforms" may ultimately prove to have been misleading. Yet, the widespread belief that the economic reforms have had a significant and positive impact on China's growth performance is borne out by the annual rate of growth since 1978. The average annual rate of growth of national income increased from 7.21 percent in 1952-1978, to 8.75 percent in 1979-1986, an increase of over 20 percent.[19]

Thus, the economic reforms have even improved upon China's past record as a growth economy. Equally important, the amplitude of the cyclical behavior of growth has been greatly reduced. In addition, the negative time trend virtually disappears when the years 1979-1986 are added on to the time series for China's rate of growth. High growth rates, then, are a major continuity of China's economic performance in the post-Mao era. A major change, however, is the turn around in the time trend and the considerable reduction in the cyclical behavior of China's economic growth.[20]

THE PATTERN OF GROWTH: STRUCTURE

Are these impacts of the economic reforms a general result that is found in all sectors alike or has there been a significant change in the structure of China's economy as well?

By the mid-1970s, the level of output was well below its production possibilities frontier, especially in the agricultural sector, as a result of poor and misguided economic policies and strategies. The decisions made at the Third Plenum singled out the agricultural sector as the starting point for the ensuing economic reforms, immediately calling for a significant increase in the prices received by producers in that sector.[21] As a result, the economic reforms in the agricultural sector have produced the most significant and dramatic systemic changes in China's economy.

Yet, even though the average annual rate of growth in the agricultural sector in 1979-1985 was more than four times its 1953-1985 average and that sector's share of national income increased from 35.4 percent in 1978 to 44.3 percent in 1983; its share fell to 41.4 percent by 1985.[22] Thus, even though the reforms in the other sectors started later and the nature of their impact on the growth rates in the non-agricultural sectors were somewhat less dramatic and more difficult to determine, annual average growth rates were higher in 1979-1985 also in the construction, transport, and commerce sectors. Only for industry was there an actual decline, from 11.1 percent in 1953-1977 to 9.1 percent in 1979-1985. This decline can be totally attributed to the reductions in output in the heavy industrial sector as the output mix in that sector was readjusted to better match the derived demand for producers' goods. In the light industrial sector, as in most other economic sectors, the economic reforms appear to have had a significant positive effect; the average annual rate of growth in the light industrial sector increasing from 9.1 percent in 1953-1977 to 12.3 percent in 1978-1985.

The impact of the economic reforms on the structure of production, therefore, was largely a reduction in the share of national income attributable to the industrial sector and an increase in the share attributable to all other sectors. This change was due to the impact of the "restructuring" of production in the economy on the demand for producers' goods and the declines in output as industries in that sector adjusted their production to meet these changes in demand. Other than that, the reforms appear to have increased the rate of growth in all sectors, including light industry. Thus, although agriculture increased its share of national income slightly at the expense of industry, the eight years of economic reform have not led to a significant restructuring of China's economy (See Table 2). While the impact of the change in priorities in favor of agriculture and light industry is recognizable, the major impact of the attempt to restructure the Chinese economy undoubtedly will be a matter of much longer-run, cumulative changes.

A most serious consequence of this major continuity from the past, despite the reforms, is revealed in the statistics for the service sectors, especially for transportation and commerce. One of the major objectives of the economic reforms has been to remove the serious bottlenecks to the flow of goods in the economy. The Chinese have tried to

Table 2
STRUCTURE OF PRODUCTION
Share of National Income (in percent)[23]

		Agr.	Ind.	Con.	Trns.	Comm.
China	1952	57.7	19.5	3.6	4.3	14.9
	1957	46.8	28.3	5.0	4.3	15.6
	1962	48.8	32.8	3.5	4.1	11.6
	1965	46.2	36.4	3.8	4.2	9.4
	1970	41.3	40.1	4.1	3.8	10.7
	1975	39.4	44.5	4.5	3.8	7.8
	1980	39.1	45.8	5.0	3.4	6.7
	1985	41.4	41.5	5.5	3.5	8.1
USSR	1984	19.9	46.0	10.7	6.0	17.4
GDR	1984	12.7	62.2	6.9	6.1	9.1
Czechoslovakia	1984	8.8	60.0	11.1	4.1	16.4
Poland	1984	17.5	49.6	11.6	5.9	13.7
Hungary	1984	18.1	46.6	11.4	8.9	13.6
Yugoslavia	1976/1980	16.4	40.5	8.3	7.4	20.4

Legend: Agr. = Agriculture; Ind. = Industry; Con. = Construction; Trns. = Transportation; Comm. = Commerce.

correct for the traditional neglect of the service sectors, a neglect found true for most Soviet-type economies. Yet, not only has the share of these sectors in China's national income declined between 1977 (13 percent) and 1985 (11.6 percent),[24] those shares remain well below the norm for the Soviet-type economies. One of the reasons for this startling statistical result of China's program of economic reform, of course, is that the statistics for national income in these Soviet-type economies (including China) exclude most consumer services and some service activities in the material producing sectors. To obtain a better estimate of the sectoral structure of economic activity in China, therefore, we can rely on statistics for employment, assuming the category "other" captures those Chinese working in service activities that are excluded from the statistics for national income (See Table 3). The statistics for the structure of employment, however, still indicate the limited extent to which the Chinese have corrected the imbalances in the structure of their economy, largely the result of the misguided economic policies pursued in the past. In other words, the economic reforms have identified and begun to cope with this problem, but have made limited or marginal gains—if any—in solving it.[25]

Table 3
STRUCTURE OF EMPLOYMENT
Share of Employment in 1984 (in percent)[26]

	Agr.	Ind.	Con.	Trans.	Comm.	Other
China	69	13	4	2	5	7
USSR	19	29	9	8	10	25
GDR	4	45	7	8	11	25
Czechoslovakia	14	38	8	7	11	22
Poland	31	29	7	8	9	17
Hungary	23	32	7	8	10	20
Yugoslavia	5	42	10	8	13	22
United States	4	22	6	6	21	41
Japan	9	26	9	6	23	27
Germany	6	34	7	6	15	32
France	8	25	8	7	16	36
Indonesia	61	8	2	2	15	12
Philippines	50	10	4	5	13	18
Thailand	68	9	2	2	9	10
Pakistan	52	15	5	5	12	11
Burma	66	9	2	3	10	10
Brazil	30	16	7	4	10	33

Legend: Agr. = Agriculture; Ind. = Industry; Con. = Construction; Trns. = Transportation; Comm. = Commerce.

The allocation of national income between consumption and accumulation was another serious imbalance in China's economy that has been identified and attacked in the program of economic reforms. As a result of their use of a Soviet-type economy and implementation of a Stalinist development strategy, the Chinese achieved one of the highest rates of savings and investment over a prolonged period for any economy in the world. These savings were largely "forced" savings and personal consumption per capita increased by only 2.1 percent a year in 1953-1977, despite the 3.7 percent a year increase in per capita national income.[27] The Soviet-type economic system, by design, assures a relatively high rate of forced savings; China's rate of accumulation already averaged 24.2 percent during the First Five-Year Plan period, 1953-1957 (See Table 4). Nonetheless, reliance upon increases in the level of investment as a major means for securing growth resulted in an increase in the rate of accumulation to as much as 33 percent during the Fourth Five-Year Plan period, 1971-1975.

The attempts to foster "consumerism" and cut the rate of investment as part of the economic reform program did succeed in reducing the rate of accumulation from 32.3 percent in 1977 to 28.3 percent in 1981,[28] i.e., during the period when output in the producers' goods sector was being modified to facilitate a restructuring of the economy. With the restoration of rapid growth in the industrial sector, especially the heavy industrial sector, and local level enterprises and units of government being able to engage in out-of-plan investment with self-provided funds, the rate of accumulation grew rapidly to regain its average level during the 1970s, increasing to 35.3 percent in 1985 (See Table 4). In fact, the rate of accumulation had averaged 29.1 percent for the period 1953-1977 as a whole, but increased to an average of 31.7 percent in the period 1978-1985.[29] Thus, rather than fostering a more balanced, consumer-oriented structure in the disposal of national income, the Chinese economy had become more accumulation-oriented as a result of the economic reforms; even by the standards of the other Soviet-type economies (See Table 4).

A much more successful structural change was the

Table 4
RATES OF ACCUMULATION OR INVESTMENT[30]
(AS A SHARE OF THE NATIONAL ECONOMY)

China (rate of accumulation)	1953-1957	24.2
	1958-1962	30.8
	1963-1965	22.7
	1966-1970	26.3
	1971-1975	33.0
	1976-1980	33.2
	1981-1985	30.8
	1985	35.3
	1986	34.3
USSR (rate of accumulation	1984	27.2
GDR (rate of accumulation)	1984	18.9
Czechoslovakia (rate of accumulation)	1984	20.1
Poland (rate of accumulation)	1984	24.7
Hungary (rate of accumulation)	1984	11.3
India (rate of gross domestic savings)	1984	22
Low-income countries (excluding China and India)	1984	7
Indonesia (rate of GDS)	1984	20
Philippines (rate of GDS)	1984	18
Thailand (rate of GDS)	1984	21
Malaysia (rate of GDS)	1984	32
Pakistan (rate of GDS)	1984	6
Burma (rate of GDS)	1984	17

attempt by the Chinese economic reformers to have China's economy better integrated into the international economic system, increasing China's foreign trade dependency. In the 1950s, China's foreign trade dependency rate was somewhat low, even for a large continental economy well endowed with resources. Nonetheless, approximately 40 percent of the machinery and equipment for China's large investment program in the 1950s came from abroad.[31] As a result of the autarkic development strategy of the Maoists, China's foreign trade dependency ratio was falling during the 1960s to an abnormally low level (See Table 5). Under the leadership of Zhou Enlai and a rehabilitated Deng Xiaoping, China's foreign trade dependency rebounded to approximately 10 percent during 1971-1977.

Table 5
FOREIGN TRADE PARTICIPATION RATIOS[32]
(Exports and Imports/National Income; in percent)

China		1953-1957	12.1
		1958-1962	10.6
		1963-1965	8.5
		1966-1970	7.1
		1971-1975	9.4
		1976-1977	10.6
		1978-1980	13.7
		1981-1985	22.2
		1986	33.1
USSR	(X+M)/GDP	1984	21.1
United States	(X+M)/GDP	1984	15.4
India	(X+M)/GDP	1984	13.8
Poland	(X+M)/GDP	1984	29.6
Hungary	(X+M)/GDP	1984	82.6
Yugoslavia	(X+M)/GDP	1984	54.8
Indonesia	(X+M)/GDP	1984	27.2
Philippines	(X+M)/GDP	1984	34.7
Thailand	(X+M)/GDP	1984	42.5
Pakistan	(X+M)/GDP	1984	30.6
Burma	(X+M)/GDP	1984	9.0

Deng's advocacy of an open economy policy had led to his expulsion from positions of power in the past, but it is the one economic reform policy he can claim as his own. As such, it was one of the first changes in economic policy introduced after his return to power, is a reform policy he has clearly endorsed continuously since his return to power, and is currently held forth as one of the two economic principles, along with four political principles, that are to be upheld by all Communist Party members. Thus, it is no surprise that China's foreign trade dependency ratio has accelerated greatly since 1977.

Due to the use of the official exchange rate to convert foreign prices into domestic prices, estimates of China's foreign trade dependency ratio are overestimates and not comparable with the estimates for other countries.[33] Even so, the open economy policy has been pursued with such intensity that China's foreign trade dependency ratio must be well above that of other large continental economies and even some small economies at China's level of economic development. The Chinese also changed their policy in regard to foreign investment to relieve the foreign exchange constraints upon the growth of their foreign trade. In 1979 through the end of 1985, China had utilized $21.787 billion (US) in foreign capital, $15.727 billion from foreign loans and $6.060 billion from direct foreign investment.[34] Had it not been for the foreign exchange constraint, China's foreign trade dependency may have increased even faster than it actually did. The Chinese were a partner in less than one percent of the international trade transactions in the world economy in 1980; by 1984, China's share was 1.4 percent. China's foreign trade is now larger than that for any Eastern European economy and larger than that for any low-income or lower middle-income country in the world.[35]

Two other very significant changes in the "structure" of China's economy are the change in ownership and the complementary change in the share of economic activity which takes place within the market sector. The official statistics for the share of the labor force employed in the state, collective, and individual sectors or the share of retail trade accounted for by each of these sectors only serve as indicators of these significant changes. For example, the statistics for employment in 1980 show that only 18 percent of the labor force was employed in the state sector (See Table 6). Before the economic reform program, however, the urban and rural collectives, which employed almost all of the remaining laborers, were really economic units directly controlled and administered by state appointed cadres. Thus, although the relative shares of employment in the different sectors had not changed very much by 1985-1986, the private and collective sectors were being restored as independent decision-making units, now recognized as not only legitimate sectors for a commodity economy at China's level of development but as necessary for achieving the transition to socialism. Thus, the development of the cooperative and individual sectors is being encouraged, especially for the provision of services.

For example, in agriculture, the commune system had been abandoned. While most peasants technically still belonged to a collective unit, basic production and income distribution decisions were made by the households under the household contract system. Some households were allowed to specialize in commodity production and market transactions as "specialized" households, while some individuals were allowed to leave farming and take up trades in the local villages or towns. As for the urban sector,

Table 6
EMPLOYMENT OF LABOR FORCE BY TYPE OF OWNERSHIP[36]
(share of total, in percent)

	State	Joint	Private	Individual	Collective[37]	
					Urban	Rural
1952	5.7	0.1	1.8	4.3	0.1	88.0
1957	8.9	1.5	-	0.4	2.7	86.5
1962	12.8	-	-	0.8	3.9	82.5
1965	13.0	-	-	0.6	4.3	82.1
1970	13.9	-	-	0.3	4.1	81.7
1975	16.8	-	-	0.1	4.6	78.5
1980	19.1	-	-	0.2	5.8	74.9
1985	18.0	0.1	-	0.9	6.7	74.3
1986	18.2	0.1	-	0.9	6.7	74.1

although the revived individual sector in cities and towns may only amount to about one percent of the labor force in 1985, it represents 4.5 million workers, compared to the 150 thousand individuals so employed in cities and towns in 1977.[38] As for the approximately 75 percent of the work force employed in the "collective" sector in both 1980 and 1986, their economic activities were much less under the control of the economic administrative bureaucracy in 1986, than was true just six years earlier.

This change is better reflected in the statistics for retail sales (See Table 7). In 1975, the state enterprises accounted for 90 percent of retail trade and much of the rest was carried out by state controlled and administered cooperatives. Now they are allowed, even urged, to compete with state enterprises so as to free up the severely strictured channels of domestic trade. Cooperatives account for a market share almost as large as that of state enterprises, while individual merchants and peasants account for one out of every five *yuan* of retail sales. Much of this retail trade, of course, is carried out in markets and most agricultural production is carried out by households responding to market prices. The state trade and supply network no longer acquires the agricultural products needed by "assigning" fixed quotas on the producer according to a plan, but now must "negotiate"

Table 7
RETAIL SALES BY TYPE AND OWNERSHIP [39]
(share of total, in percent)

	State	Joint	Individual	Collective	Peasants[40]
1952	34.4	0.4	60.9	-	4.3
1957	62.1	16.0	2.7	16.4	2.8
1962	82.0	-	2.4	11.2	4.5
1965	83.3	-	1.9	12.9	1.9
1975	90.2	-	0.1	7.7	2.0
1980	84.0	-	0.7	12.1	3.2
1985	40.4	0.3	15.4	37.2	6.8

with the peasants to obtain the quantities of output required; the peasants agreeing to deliver a fixed quantity of output at an agreed price at harvest time in an "advanced purchase contract." State enterprises still produce to meet their assigned quotas under the plan, transferring that output to the state at administered prices and buying their inputs also at administered prices. But even state enterprises now produce above quota output for sale on the market, also using the market to acquire inputs to produce that output. Almost every industrial commodity in China today has a market price (the marginal price), a negotiated price (usually part of a barter transaction with other enterprises), and an administered price (the state's purchase and wholesale prices). To what extent China's Soviet-type economy (which always included a planned sector, a barter sector, and a market sector) has been changed to become a "mixed economy" with the flow of resources and commodities in the economy being determined approximately equally by the planned targets and market forces is difficult to determine with any degree of precision. One survey of 429 industrial enterprises in the state sector in 1985 found that slightly less than one-third of the enterprises secured over 40 percent of their input in the market sector, while slightly more than half of them marketed over 40 percent of their output.[41] Another source cites the decline in the number of commodities handled by the state's unified distribution system (the planned allocation of input), from 256 commodities in 1978 to only 20 in 1986, and the decline of the state's unilateral budget grants for local level investments, from 76.6 percent of total investment at the provincial and municipal level in 1978 to 31.6 percent of the total in 1986. Finally, this same source claims that commodities traded according to prices fixed by the state account for 35 percent of the trade in agricultural products, 45 percent for industrial consumer goods, and 60 percent for industrial raw materials.[42] These statistics do indicate that the Chinese economy has changed from a Soviet-type economy, totally dominated by state ownership and control with central planning guiding the allocation of resources and commodities, to a mixed economy with a much greater private and cooperative sector and a larger role played by markets.

While the debate continues over how far those changes have gone, few would deny that very significant changes have already occurred and have done so as a result of the economic reform program. In fact, a reasonable hypothesis to make on the basis of the available evidence would be that, with the exception of Yugoslavia, the Chinese have moved further along the road to market socialism as a result of their economic reforms than any other Soviet-type economy. How much further along that road they are likely to get in the future is a question we will return to in the conclusion.

SOURCES OF GROWTH

In an economy as large as China's, the extensive development program in the three decades after 1949 obviously involved productivity gains and technological innovation in some sectors over varying periods of time.[43] Yet, most attempts—some crude and some very sophisticated—to obtain macro-economic estimates for the sources of growth in China's economy in 1952 through the 1970s all show the same result; China's economic development over this period is a classic example of extensive economic growth: a rapid increase in output obtained by means of a proportionately rapid increase in input and production facilities.[44]

Perhaps the crudest attempt to estimate the sources of growth for China's post-1949 economic development is my own.[45] In that estimate, a single measure for the increase in inputs was obtained by adding together the increase in labor and the increase in the book value of fixed assets according to a fixed set of weights, comparing this increase in inputs with the increase in national income, and assuming that any difference between the rate of growth in inputs and outputs (which obviously could be due to a host of factors) was totally due to the change in productivity of the inputs. For the period 1953-1980, our results showed that when labor is given a weight of 40 percent (capital, a weight of 60 percent) the average annual change in total factor productivity was -1.8 percent; when capital is given a weight of 40 percent, total factor productivity remains almost constant over this long period.

The most dynamic growth sector in China after 1949 was industry, but even if we restrict the analysis of the sources of growth to the industrial sector, estimates produced by the World Bank do not show strong productivity growth.[46] When labor is given a weight of 40 percent, total factor productivity in industry increases by 0.6 percent a year in 1952-1982; when capital is given a weight of 40 percent, the average annual rate of increase becomes 1.5 percent. A much more sophisticated methodology requires the specification of an equation for the production function and the use of econometric analysis to estimate the various parameters of that function. This approach probably is more sophisticated than is warranted by the data available for the analysis. Yet, Gregory Chow has used this method to analyze the available data for outputs and inputs in the industrial sector, obtaining remarkably consistent and rational results; concluding that industrial output in 1952-1981 increased in China mainly as a result of the increase of capital assets, rather than an improvement in productivity or technology.[47]

Gerald Meier claims that only half the explanation of economic growth in the successfully developing countries in Asia can be explained by the increase in inputs over time, while half of the growth "must be attributed to technical progress, improved quality of labor, and better management that combines the inputs more effectively."[48] A major objective of China's economic reform program is to shift from extensive to intensive economic growth, i.e., to reduce the share of output growth due to the mere increase in the level of inputs and creation of new production facilities, and to increase the share of growth due to productivity increases and technological change.

Perhaps an attempt to evaluate the extent to which the economic reforms have been successful in achieving this objective is premature.[49] Nonetheless, there is some evidence that suggests a major continuity in China's economic development is the dominance of extensive, as against intensive, growth. For example, the most successful results of the economic reforms have been observed in agriculture. Yet, Justin Yifu Lin has analyzed the sources of growth in the agricultural sector after the introduction of the household responsibility system and his results show that reform policies can account for about 45 percent of the growth in agricultural output in 1980-1983, good weather can take credit for five percent, while the increase of inputs still accounts for half of the increase in outputs.[50] More important, the increase in productivity due to the economic reforms occur mostly in the initial years the reforms were introduced and the importance of this source of growth decreases over time.

After the 1980-1983 period, i.e., 1984-1986, crop production accounted for only 11.2 percent of the total increase of agricultural output and forestry, animal husbandry, and fishing combined for only 16.6 percent of the total.[51] Over 70 percent of the growth in agricultural production after 1983 was accounted for by the increase in sideline activities. Village-run industrial enterprises accounted for 84 percent of these sideline activities in 1986;

it is the rapid growth in these industrial enterprises, of course, that explains a large part of the rapid growth of investment that has occurred outside the plan and outside the state's budget.[52] Thus, after 1983, a significant source of growth in the agricultural sector has been due to the very rapid growth of industrial production in village-run industrial enterprises. The source of growth in that sub-sector certainly must be largely explained by the rapid growth of self-provided investment in the creation of these industrial enterprises (fixed assets in township and village enterprises—not just industrial—increased by 58 percent between 1983 and the end of 1985)[53] and the equally rapid growth in their share of the rural labor force (employment in township and village enterprises had grown to a total of 44 million by 1986.)[54] While the absorption of surplus agricultural labor in these rural enterprises served to reduce disguised unemployment in the agricultural sector and productivity gains and innovation obviously made a contribution to the remarkable record of growth in the agricultural sector since 1977, that contribution may well have been largely concentrated in a one-time shift in factor productivity, with increases in inputs still explaining well over half of the increases in output, especially in the past few years.

In industry, the rate of increase in the net value of fixed assets declined to less than the rate of increase in the gross value of industry in 1978-1985, whereas it had been higher in the period 1953-1977 (See Table 8). Except for the early 1970s, the increase in employment has always been less than the increase in output in industry. Thus, the rate of growth of total factor productivity in industry was not only greater in 1978-1985 than in 1953-1977, but also accounted for a significantly larger share of the growth in output. According to the estimates in Table 8, productivity growth accounted for between 16 and 23 percent of the rate of output growth in 1953-1957; 39 to 51 percent in 1978-1985. If accurate, these estimates would be a most significant argument for the successful impact of the economic reform program on China's economic growth in the post-1977 period.

Table 8
SOURCES OF GROWTH: STATE OWNED INDUSTRY
(annual average rate of growth, in percent)[55]

	Gross Value of Output	Net Value of Fixed Assets	Employment	Total Factor Productivity	
				(1)	(2)
1953-1957	24.2	24.4	8.0	6.4	9.6
1958-1965	14.6	16.2	6.5	2.3	4.2
1966-1970	11.4	5.6	9.6	4.2	3.4
1971-1977	4.8	7.4	6.3	-2.2	-1.9
1953-1977	12.9	13.1	7.4	2.1	3.0
1978-1985	8.8	7.5	2.2	3.4	4.5

Weights in total inputs: (1) Employment, 40 percent; fixed assets, 60 percent. (2) Employment, 60 percent; fixed assets, 40 percent.

Unfortunately, due to the lack of statistics for the net value of output on an annual basis, the rate of growth of the gross value of output had to be used. Material consumption per unit of output increased between 1975 and 1985 (See Table 9). In other words, the net value of output could be growing slower than the gross value of output inasmuch as current costs per unit of output increased over this period. On the other hand, comparable data are available only for the two years 1980 and 1985; they show the growth in net value of output was slightly higher than the growth in gross value of output in state-owned industry over this period.[56] Thus, it seems unlikely that our use of the rates of growth for the gross value of output, rather than the net value of output, results in a serious bias in our estimates.

This seemingly technical problem in obtaining estimates for changes in productivity and efficiency in Chinese industry is of some importance, however. Although the estimates in Table 8 do indicate that the growth of output per unit of capital and labor after 1977 does explain a larger share of total industrial growth than was true in the pre-reform period, i.e., the economic reforms were making a difference in terms of increases in total factor productivity; the statistics in Table 9 and Table 10, also show that current inputs were not being used more efficiently in combination with the inputs of capital and labor. Over the period 1977-1985, material consumption per unit of output did not decline, costs per unit of sales increased; with prices increasing, the amount of profit and taxes per unit of capital increased only slowly. Thus, the evidence is somewhat mixed and it may, indeed, be premature to pass judgment upon the success of the reform attempts to achieve such a fundamental change in direction, i.e., to make increases in productivity and efficiency a major source of economic growth.

What can be said with some certainty, however, is that increases in the level of investment remain a very important explanation of economic growth in China; there is limited

Table 9
PROPORTION OF MATERIAL CONSUMPTION TO GROSS OUTPUT VALUE BY SECTOR[57]

	Agr.	Ind.	Con.	Trns.	Comm.
1952	26.2	66.9	63.2	28.6	22.1
1957	20.9	63.5	61.9	35.0	24.1
1962	24.0	67.1	56.8	38.7	33.1
1965	23.0	64.0	69.9	36.3	32.7
1970	24.9	62.9	71.2	36.8	25.6
1975	26.7	64.4	74.1	40.0	37.8
1980	33.8	65.5	75.9	49.0	43.9
1985	38.3	67.7	76.9	46.6	38.8

Legend: Agr. = Agriculture, Ind. = Industry, Con. = Construction, Trns.= Transportation, Comm. = Commerce

Table 10
PRINCIPLE FINANCIAL ITEMS OF STATE OWNED ENTERPRISES[58]

	(1)	(2)	(3)	(4)
1952	134	14.1	37.0	69.9
1957	139	16.9	47.7	69.1
1962	71	12.5	20.5	76.5
1965	98	21.3	39.8	69.0
1970	117	18.2	45.7	69.6
1977	99	13.5	31.5	74.1
1980	101	15.5	35.9	73.7
1980		(23.3)		
1985	95	(22.4)	33.5	n.a.

Legend: 1. Gross output value per rmb100 of original value of fixed assets. 2. Profits per rmb100 of gross output value. Figures in parentheses are for profits and taxes per rmb100 of gross output value. 3. Profits and taxes per rmb100 of net value of fixed assets. 4. Costs per rmb100 of revenue from sales.

statistical evidence, in the first half of the 1980s at least, of a significant breakthrough with regard to improved efficiency and productivity in Chinese industry.[59] Rather than a major change in the role played by investment and increases in inputs as a source of economic growth, the economic reform program has brought about a major change in the sources of financing that investment. As a result of the decentralization process, investment in state-owned units was only two-thirds of the total investment in fixed assets in 1985; investments financed by allocations within the state budget accounted for only 17.7 percent of the total.[60] Furthermore, about half the state budget is now accounted for by local revenues and expenditures, while investment financed by domestic loans is now larger than that financed by the state budget. Investment financed by "self-raised funds" (retained earnings, depreciation accounts, local revenues, bond sales, etc.) accounted for almost two-thirds of all investments in fixed assets in 1985. Thus, while investment appears to have remained as a major source of growth in China's economy, the transition to more "voluntary" means for financing that investment, compared with the almost exclusive reliance on "forced" savings in the past, can be considered as a major result of the economic reforms in the post-1977 period.

THE QUALITY OF GROWTH

One of the consequences of the extensive development strategy pursued before 1978 was the failure of the individual to obtain a significant share of the benefits of the resulting higher levels of total output. Part of the increase in inputs came from an increase in the labor force participation rate and a reduction in unemployment. Thus, even with a constant income per worker, the resulting decline in the dependents per worker would lead to an increase in per capita incomes. To illustrate this, the increase in the labor force participation ratio from 36.1 percent in 1952 to 47.7 percent in 1985 and the decline in the unemployment rate from 13.2 percent to 1.8 percent over the same period yields a 50 percent increase in per capita income with no increase in the annual incomes of those actually working.[61] These changes must account for a considerable share of the increases in per capita incomes, at least before 1978; but while per capita income was increasing at six percent annually in 1953-1978, real per capita consumption was increasing by only two percent annually.[62] In regard to food supplies, there were actual declines in per capita consumption of grain and edible vegetable oils and almost negligible increases in the per capita consumption of meat, sugar, and cloth between the mid-1950s and the mid-1970s (See Table 11). Finally, over 25 years after the founding of the People's Republic, urban residents had less than five square meters of living space and less than a hundred *yuan* of savings per capita (See Table 12). Housing conditions for rural residents were somewhat better, but still less than 10 square meters per capita; peasants' savings were almost depleted, amounting to less than 10 *yuan* per capita.

A major objective of the economic reforms has been to create material incentives for obtaining the local initiative and efforts needed to achieve greater efficiency and productivity. As a result, one of the greatest economic

Table 11
PER CAPITA CONSUMPTION OF CONSUMER GOODS[63]

	Grain (kg.)	Edible Vegetable Oil (kg).	Meat (kg.)	Sugar (kg.)	Cloth (m.)
1952	197.7	2.1	7.3	0.9	5.7
1957	203.1	2.4	6.7	1.5	6.8
1962	164.6	1.1	3.4	1.6	3.7
1965	182.8	1.7	7.7	1.7	6.2
1970	187.2	1.6	7.2	2.1	8.1
1975	190.5	1.7	8.7	2.3	7.6
1980	213.8	2.3	12.8	3.8	10.0
1985	254.4	5.1	16.9	5.6	11.7

Table 12
PER CAPITA HOUSING SPACE AND SAVINGS[64]

	Housing (Living space in sq. meters)		Savings (in current rmb)	
	Urban	Rural	Urban	Rural
1978	4.2	8.1	89.8	7.0
1979	4.4	8.4	109.5	9.9
1980	5.0	9.4	147.6	14.7
1981	5.3	10.2	175.5	21.2
1982	5.6	10.7	211.4	28.4
1983	5.9	11.6	237.3	40.8
1984	6.3	13.6	235.3	62.2
1985	6.7	14.7	276.6	85.2

consequences directly attributable to the economic reform program is the dramatic increase in per capita income. Offsetting some of these gains was another result of the economic reforms, i.e., inflation, but urban workers were given subsidies to partially offset the higher prices for foodstuffs and necessities. In current prices, per capita consumption of the peasants increased at an average annual rate of 9.9 percent annually in 1975-1986, while that for non-agricultural residents increased by 9.3 percent annually. Despite the inflationary price rises over this same period, real per capita consumption of the peasants increased by 6.7 percent annually between 1975 and 1986, while the rate of growth for non-agricultural residents—despite the subsidies—was somewhat lower at 5.5 percent annually (See Table 13).

Perhaps of greater significance to the individual consumer was the increase in the per capita consumption of basic foodstuffs. Between 1975 and 1985, per capita grain consumption not only increased at the rate of 2.9 percent annually, the increase in the level of per capita consumption of grain included a significant shift in consumption from the poorer grains to the richer grains. The annual average

Table 13
PER-CAPITA LEVELS OF CONSUMPTION[65]
(Index numbers, 1952 = 100)

	Peasants		Non-agricultural Residents	
	Current Prices	Constant Prices	Current Prices	Constant Prices
1952	100	100	100	100
1957	127	117	139	126
1962	142	99	153	97
1965	161	125	160	137
1970	184	142	176	152
1975	202	151	219	187
1980	279	185	316	238
1985	523	299	509	313
1986	568	308	584	336

increase in the per capita consumption of edible vegetable oil, meat, sugar, and cloth over this same decade was 11.6 percent, 6.9 percent, 9.3 percent, and 4.4 percent, respectively (See Table 11).

The differential impact of the economic reforms tended to significantly narrow the ratio of real per capita consumption of the peasants to that of the non-agricultural residents (See Table 14). Although this ratio remained higher than was true for the 1950s and it would take a long time for the peasants' real per capita consumption level to approach that of the urban worker,[66] the economic reforms did lead to two dramatic changes that were readily observable to all. The peasants have used their new freedoms and increased incomes to rapidly increase their available living space; by 8.9 percent annually between 1978 and 1985 (See Table 12). Although the housing space per urban dweller has increased more rapidly after 1977 than during the pre-reform period (by 6.9 percent annually), by 1985 it was less than half that of the rural population, despite the smaller size of the typical urban family. Housing is now said to be the worst problem facing the typical resident of Shanghai today.[67] Another reason for the urban workers feeling they are falling behind is that not only has the

Table 14
RATIO OF PEASANTS TO NON-AGRICULTURAL RESIDENTS PER-CAPITA CONSUMPTION[68]

	Current Prices	Comparable Prices
1952	1:2.4	1:2.4
1957	1:2.6	1:2.5
1962	1:2.6	1:2.3
1965	1:2.4	1:2.3
1970	1:2.3	1:2.6
1975	1:2.6	1:2.9
1980	1:2.7	1:3.1
1985	1:2.3	1:2.5
1986	1:2.5	1:2.6

peasants' per capita real consumption increased faster than the urban dweller, but their available liquid assets are growing much faster as well.[69] In 1978, per capita urban savings were more than ten times that of the typical rural resident, the latter's savings averaging only seven *yuan*. Between 1978 and 1985, however, the average per capita savings of the rural resident had grown by over 40 percent annually and per capita savings in the urban sector was only slightly more than three times those of the rural resident.[70]

The two major objectives of the economic reforms, i.e. to achieve greater efficiency and productivity and a greater standard of living, have both led to a decentralization of control over the allocation of resources and income. Thus, local areas, enterprises, and individuals now have greater freedom to mobilize their own resources and produce for the market and material incentives now relate income earned with skill and effort of work performed. Received theory tells us that these reforms would lead to a worsening of the distribution of income, at least regionally; better endowed areas in terms of resources, infrastructure, and markets would be better able to take advantage of the reforms.[71] Yet, if the economy in the mid-1970s was operating well within its production possibilities frontier it should be possible, for a period of time at least, for everyone to become better off as a result of the economic reforms, although the result is indeterminate insofar as the class-size distribution of personal income for the economy as a whole is concerned.

Despite the flood of data being published by the Chinese in recent years, time series and cross-section micro-level data for household incomes that can be used to derive class-size distribution of income statistics for China over the past decade or longer are not available. Statistics published for the rural and urban sample surveys carried out by the Chinese do indicate the economic reforms have made everyone (or almost everyone) better off in an absolute sense (See Table 15). Whether the rich are getting richer faster than the poor are getting less poor is an important question and the answer to that question may have important consequences on the popular support for the reforms.[72] Nonetheless, before 1977 we observed the attempt to achieve greater equality of income by making everyone "poor," whether that attempt actually made the distribution of income more equitable or not is debatable.[73] When the average level of per capita consumption increased slowly, researchers tried to find evidence for the assumed offsetting benefit of a more equitable distribution about the mean. Now that the average per capita consumption has begun to grow very rapidly after 1977, we search for the

Table 15
DISTRIBUTION OF INCOME
(in percent of households in sector)[74]

Urban Households with annual per capita income available:	1981	1982	1983	1984	1985
RMB240 and below	2.1	0.9	0.6		
RMB240-300	5.5	3.7	3.0		
RMB300-420	31.8	25.6	20.31	10.5	4.3
RMB420-600	42.3	45.4	46.6	38.9	22.2
RMB600-720	11.9	14.2	16.4	22.7	20.9
RMB720 and over	6.5	10.2	13.1	26.25	51.6

Peasant Households per capita income:	1981	1982	1983	1984	1985
RMB100 and below	4.7	2.7	1.4	0.8	1.0
RMB100-150	14.9	8.1	6.2	3.8	3.4
RMB150-200	23.0	16.0	13.1	9.4	7.9
RMB200-300	34.8	37.0	32.9	29.2	25.6
RMB300-400	14.4	20.8	22.9	24.5	24.0
RMB400-500	5.0	8.7	11.6	14.1	15.8
RMB500 and over	3.2	6.7	11.9	18.2	22.3

evidence of the assumed cost of that growth, i.e., a reduced equity in the distribution of income. This merely reflects the commonly accepted assumption that there is an inverse relationship between the rate of growth in the level of average per capita income and the distribution about the mean, an assumption that is increasingly being challenged.[75] China's contemporary experience may well be the latest, yet most significant, evidence that the relationship between the average and its distribution can be a direct one, rather than inevitably being an inverse one.[76]

Whichever side of the argument the evidence turns out to support in regard to the changes in relative incomes, the statistics in Table 15 clearly indicate that the absolute level of real income for a sizable portion of the poorer households has increased significantly as a result of the economic reforms. For those who accept the "needs" test as the most meaningful way to judge the results of economic development programs, this statistical result of the economic reforms in China would be one of the most important of all and would point to a major discontinuity, for the better, in China's modern economic development efforts.

CONCLUSION: CONTINUITY AND CHANGE

Our survey of the more readily available macro-statistics for recent economic developments in China and our comparison of those results with those in earlier periods and in other countries represents a sort of stock-taking at the end of the first decade of the Deng/Zhao leadership's attempts to implement a major program of economic reforms and to create a new mixed and open economy, or a "socialist economic system with Chinese characteristics." In addition to identifying the extent to which that attempt has achieved some successes, we believe this survey also helps us identify the economic problems that remain to be solved or that have been created as a result of the economic reforms. For example, we found that the economic reforms may be associated with a slight increase in what had already been a very high rate of growth in the pre-reform period and that the rate of growth may have been made more stable as a result of the economic reforms. At the same time, a significant explanation for the higher rate of growth was to be found in the agricultural sector and much of that growth was either explainable by one-time shifts in the production function or by very rapid increases in non-crop activities from a very low base. Has China's long-run growth path been accelerated as a result of the economic reforms or has it experienced a discontinuous jump to a higher level of output, most likely to return to a pattern of growth similar to that in the pre-1978 period?

We also found that despite strenuous efforts to lower the rate of accumulation and investment and to restructure the economy, the rate of accumulation has declined only slightly and the structure of the economy, after an initial change in favor of agriculture and light industry, has also been modified only slightly. While productivity increases explain a larger share of growth, the record of efficiency in the use of current inputs and the profitability of economic units remains very poor. Investment remains a significant source of growth in China's economy and the attempt to change from extensive to intensive growth is progressing very slowly. In other words, there is sufficient evidence to indicate that the Chinese economy is still operating well within its production possibilities frontier, especially if judged from the standpoint of modern technology.

On the other hand, significant discontinuities from the past have occurred and these changes are not likely to be reversed. One of the key features of the leadership's reform program is the "Open Door" policy, with the result that China's foreign trade ratio has been increased significantly, to the point where it is unique for a large, well-endowed, continental economy at China's level of development. Foreign borrowing and investment are being actively promoted by the Chinese, while the transfer of Western technology and technical training now plays a significant role in China's economic development process. A second important discontinuity readily observed is the extent to which the population has shared in economic growth, i.e, has obtained higher standards of living.

Both of these new features of China's development strategy can be said to have benefited China's economic growth and welfare. Yet, both features also have raised serious problems for China's leadership. The Open Door policy has given rise to opposition from those who oppose the forces of "bourgeois liberalism and corruption" that has also come through the open door. This anti-Westernization sentiment has deep roots among the population in China. As for consumerism and material incentives, after a generation of basically egalitarian policies and ideological arguments, ostentatious shows of wealth and unequal incomes are creating hostility among a significant share of the population that, despite the increase in the absolute level of their income, believe they are being left behind. In addition, the dramatic increase in the quantity and variety of consumer goods available has created rising expectations that will be difficult to satisfy.

Another significant change in China's development strategy is the shift in the sources of financing investment from "forced" savings to a variety of sources that are much more "voluntary." The problem with this reform, however, is that it has been too successful, generating an excessive level of investment that is outside the "state's plan," i.e., investments in local, small-scale, duplicative enterprises throughout the economy. Any program of rationalization for Chinese industry would result in a smaller number of producers producing on a larger scale, with much greater specialization in the product produced. Having opened Pandora's box to allow for more voluntary savings and investment at the local level, however, the central authorities will find it difficult to step in and either close down these units or force them to face competitive market forces by breaking down the barriers to inter-regional transportation and trade.

Finally, the reforms also have brought about inflationary price increases, budget deficits, balance of payments problems, and open or disguised unemployment. In the past, the Chinese held forth their record of price stability, a balanced budget, the lack of foreign dependency and debt, and full employment as proof of the superiority of socialism. Despite the many beneficial results of their economic reform program, however, these latter results would indicate that they are now becoming more like the capitalist economies insofar as the problems of price stability, budget deficits, balance of payments problems, and unemployment are concerned.

The Chinese economic reforms, of course, have a long way to go before the Chinese economy can be said to resemble a market economy. Yet, our examination of the macro-economic statistics does detect a significant shift taking place in the Chinese economy; a shift away from a Soviet-type economy toward that of a socialist market economy. Yet, here too, their very successes have created a dilemma they must resolve, i.e., another Pandora's box has been opened, resulting in an unstable mix between the state and cooperative/private enterprise sectors and the plan and market sectors. While a Marxist may have faith that this dialectic will eventually yield a new synthesis, how far can the collective and/or private enterprise sectors be allowed to grow before they challenge the dominance of the state-owned sector and how far can market forces be allowed to expand their influence over the allocation of resources and distribution of goods and incomes before they challenge the control of the plan? Will the present pace of erosion of the state-owned sector and centrally planned economy continue and, if so, will the Chinese economy simply become a collective/private enterprise, market economy controlled and administered by local bureaucrats?

The political leadership in China has shown great political will and courage in carrying out the economic reforms, but very important questions remain unresolved; price reform, ownership and management rights, bureaucratic controls and interference in the economy, the role of the party and party cadre in the operation of the economy, etc. In other words, the party has yet to define more precisely exactly what is meant by "socialism with Chinese characteristics," assuming they desire a consistent, feasible, and efficient economic system and development strategy. The failure to do so, of course, will threaten the very favorable trends in the macro-economic statistics for China's economic performance that they have achieved thus far.

NOTES

1. "A new historic era in China began in December 1978. The signpost of this new era is the Third Plenum of the Eleventh Central Committee, marking not only a decisive break with those ideological and political lines of Mao which culminated in the Cultural Revolution, but also signaling the end of nearly three-quarters of a century of revolutionary ferment and upheavals...thus reversing the fundamental trend of political developments since the May Fourth period of 1917-1921." Tang Tsou, "Political Change and Reform: The Middle Course," in Norton Ginsburg and Bernard A. Lalor, eds., *China: The 80s Era* (Boulder, Colorado: Westview Press, 1984), p. 27.
2. See, for example, Robert F. Dernberger, "The Chinese Search for the Path of Self-Sustained Growth in the 1980s: An Assessment," in Joint Economic Committee, US Congress, "China Under The Four Modernizations," Vol. 1 of *China's Economy Looks Toward the Year 2000*, 2 vols. (Washington, D.C.: US Government Printing Office, 1982), pp. 19-76. This review is updated by the papers reporting on specific sectors in the latest edition in the same series; Joint Economic Committee, US Congress, *China's Economy Looks Toward the Year 2000*, Vol. 1, *The Four Modernizations* and Vol. 2, *Economic Openness in Modernizing China* (Washington, D.C.: US Government Printing Office, 1986). A very good source for detailed reporting on the various economic reform experiments and their implementation is to be found in *China Newsletter*, a bimonthly, published by JETRO (Japan External Trade Organization) in Tokyo.
3. For a description of the traditional Soviet-type economic system, see Robert F. Dernberger, "The State-Planned, Centralized System: China, North Korea, Vietnam," in Robert A. Scalapino, Seizaburo Sato, and Jusuf Wanandi, eds., *Asian Economic Development—Present and Future* (Berkeley, California: Institute of East Asian Studies, University of California, 1985), pp. 13-42. For the detailed description of a market socialist system by those who originally proposed it, see Oskar Langer and Fred M. Taylor, *On The Economic Theory of Socialism* (New York: McGraw-Hill Book Company, 1964).
4. Janos Kornai, "The Hungarian Reform Process," *Journal of*

Economic Literature Vol. XXIV, No. 4, (December, 1986), pp. 1687-1737.
5. *Ibid.*, p. 1734.
6. There are two very intense scholarly debates that are raised by the simple statement made in the text here. The traditional interpretation of Soviet economic history is that the very important "industrialization" debate between the advocates of a big-push development strategy and balanced growth was mediated by Stalin and resolved by his decision to create the Soviet-type economic system and to use that system to implement the big-push development strategy. This view is questioned by James Millar and others, who argue that the Soviet-type economic system was created by lower level cadres who were coping with real problems as they occurred. Once created, of course, Stalin claimed credit for having advocated it as the correct solution to the political and economic debates, while most of the participants in the debates—on both sides—were executed.

   In addition, the simplistic interpretation of how that system works—the mobilization of resources in the budget and its allocation to investment in priority sectors in heavy industry—has also argued that the resources mobilized were from agriculture and the Stalinist big-push development strategy is really a clear case of exploiting agriculture for the sake of industrialization. Shigeru Ishikawa and others (including James Millar) argue that we need to look at relative terms of trade, urban wages, etc.—not just the flows in the budget—to determine the total net flows of resources throughout the economy. In doing so, these observers claim that the net flow of resources for industrialization were actually provided within the urban-industrial sector itself; by the industrial workers, etc. For my own analysis of how the Soviet-type economic system was adopted and used to achieve high growth rates in China after 1949, see Robert F. Dernberger, "Financing China's Development: Needs, Sources and Prospects," in Robert F. Dernberger and Richard S. Eckaus, *Financing Asian Development: China and India* for the Asia Society, (New York: University Press of America, 1988), pp. 12-68.
7. Recognizing the many differences in the degree to which relative prices in the various countries reflect true scarcity pricing and the many differences as to what is included as output

in the estimates, several researchers have spent a good amount of effort to derive a more valid comparison between various countries' levels of economic activity and standards of living. To repeat or summarize those efforts here is both impossible and unnecessary. Almost all such studies conclude that the statistics for the socialist countries result in higher rates of growth, than if true scarcity prices and Western national income accounting methods were to be used. While I believe that removing this upward bias would not change the argument being made in this paper concerning China's comparative record of economic growth, others may disagree. Unfortunately, detailed statistical analysis and evidence is most unlikely to resolve this debate and so none is provided here.
8. Herbert S. Levine, "Possible Causes of the Deterioration of Soviet Productivity Growth in the Period 1976-1980," in Joint Economic Committee, US Congress, *Soviet Economy in the 1980s: Problems and Prospects*, Part 1 (Washington, D.C.: US Government Printing Office, 1982), p. 154.
9. Thad P. Alton, "East European GNPs: Origins of Product, Final Uses, Rates of Growth, and International Comparisons," in Joint Economic Committee, US Congress, "East European Economies: Slow Growth in the 1980s," Vol. 1 of *Economic Performance and Policy* (Washington, D.C.: US Government Printing Office, 1985), pp. 120-121.
10. *Ibid.*
11. *Ibid.*
12. *Ibid.*
13. *Ibid.*
14. *Ibid.*
15. Estimated from index of national income (1952=100) in "comparable prices." State Statistical Bureau, People's Republic of China, *Statistical Yearbook of China, 1986* (Hong Kong: Economic Information Agency, 1986), p. 41.
16. World Bank, *World Development Report, 1986* (New York: Oxford University Press, 1986), pp. 182-183.
17. The explanation for this is that the high rates of savings and investment, the fundamental source of high growth in the Soviet-type economy, are unstable. Furthermore, the original shift to high rates of savings and investment yields higher growth rates in the early years, but the advantage from adopting this economic system and development strategy is eroded over

time as inefficiencies of the system lead to higher and higher capital-output ratios.
18. Estimates are based on time-series of annual growth rates derived from the index of national income in "comparable" prices (1952=100) in *State Statistical Bureau, Statistical Yearbook 1986, loc. cit.*
19. *Ibid.* Indices for national income in more recent years, including 1986, that revise the earlier released figures are from State Statistical Bureau, People's Republic of China, *Zhongguo Tongji Zhaiyao, 1987* (Beijing: Chinese Statistics Publishers, 1987), p. 7.
20. The standard error and the coefficient of variation declined by three-fourths, while the moment coefficients of skewness and kurtosis both declined by over one-half.
21. The higher prices paid to producers depended upon not only the successive price increases announced for quota and above quota deliveries for the individual products in 1979, 1980 and 1981, but the share of output self-consumed, sold at the quota price, the above quota price, and on the market. No matter, the effective price paid the producers for their output was increased significantly. For example, in the first round of official price increases alone (March 1979), the quota prices for grain, oil-seeds, cotton, and hogs were increased by 20, 25, 15, and 26 percent, respectively; while the above quota price was set above the 1978 price by an increase of 80, 88, 50, and an unknown amount, respectively. Price increases in 18 other categories of agricultural products ranged from 20 to 50 percent. See Frederic M. Surls and Francis C. Tuan, "China's Agriculture in the Eighties," in Joint Economic Committee, US Congress, "China Under The Four Modernizations," Vol. 1 of *China's Economic Looks Toward the Year 2,000*, 2 vols. (Washington, D.C.: US Government Printing Office, 1982), pp. 428-429.
22. Statistics in this paragraph are from the absolute level of national income originating in each of the sectors or the sector's share in national income, all in current prices, from the State Statistical Bureau, *Statistical Yearbook 1986, op. cit.*, pp. 40, 43. Rates of growth for output in the light industrial sector presented at the end of this paragraph are estimated from the index of gross industrial output value (1952=100) based on comparable prices from the State Statistical Bureau, *Statistical Yearbook 1986, op. cit.*, p. 225.
23. *Ibid.*, pp. 43, 715.
24. *Ibid.*, p. 43.

25. These statistics for the service sector may properly reflect the development in the several decades during which this sector was neglected and even restricted, as well as the large job the Chinese face in restoring this sector to its "normal" role in an economy at China's present level of development. By relying on the "official" statistics for the relative shares of output and employment in the material goods-producing sectors, however, these statistics may fail to show the considerable increase in the "absolute" size of the services sector as a result of the economic reform program. According to recently released statistics for total employment of the entire labor force in the agricultural, industrial, and services sectors, the labor force in the services sector doubled in 1978-1986, increasing by more than 40 million workers, and increased its share in the total labor force from 11 percent in 1978 to 16.8 percent in 1986. State Statistical Bureau, *Zhongguo 1987, op. cit.* p. 19. These statistics for the entire labor force appear to be more comprehensive and complete than those for national income and employment in the five material producing sectors, plus "other," i.e., these statistics may be derived from the census data rather than the economic statistics collected by the State Statistical Bureau. In other words, to the extent that much service sector activity is to be found in the small-scale, private, market sector and that the State Statistical Bureau has to rely on sample statistics to capture this activity in the "official" statistics, the "official" statistics may well fail to reflect the success the economic reform program is having in reviving service sector activity.
26. *State Statistical Bureau, Statistical Yearbook 1986, op. cit.*, pp. 43, 714.
27. *Ibid.*, p. 714; estimated from the annual total per capita national income on p. 40, total annual personal consumption on p. 53, and total year end population on p. 71.
28. *Ibid.*
29. Estimated from the statistics for the annual absolute levels of national income and accumulation, in current prices, *ibid.*
30. *Ibid.*, pp. 49, 715 and *Zhongguo 1986, op. cit.*, p. 7; World Bank, *op. cit.*, pp. 188-189.
31. For the extent to which China depended upon imports for their investment program after 1949, see Robert F. Dernberger, "Economic Development and Modernization in China: The Attempt to Limit Dependence on the Transfer of Modern Industrial Technology from Abroad and to Control its

Corruption of the Maoist Model," in Frederic J. Fleron., ed., *Technology and Communist Culture* (New York: Praeger Publishers, 1977), pp. 224-264.
32. China figures are estimated from annual absolute values of national income and exports and imports, in current prices in State Statistical Bureau, *Statistical Yearbook 1986, op. cit.*, pp. 40, 481 and *Zhongguo 1986, op. cit.*, pp. 7, 89. USSR: *op. cit.*, pp. 715, 730. Other countries: World Bank, *op. cit.*, pp. 184-185, 196-197.
33. After 1980, foreign trade data was issued by the Chinese Customs and was based on foreign prices, whether reported in US dollars or in *yuan*. National income data is based on domestic prices and reported in *yuan*. In other words, China's foreign trade data and domestic economic activity data are based on incomparable prices and the use of the "official" exchange rate to convert the foreign trade data into *yuan* values or convert the national income data into US dollars to estimate China's foreign trade dependency ratio results in a serious overestimate of the true foreign trade dependency ratio. If the rate of exchange established on the market in Shanghai (The Shanghai Foreign Exchange Adjustment Center, admittedly a very special and thin foreign exchange market), where joint-ventures with a surplus of foreign exchange can sell foreign exchange to those joint-ventures with a deficit in their foreign exchange account, can be accepted as a better indication of the true purchasing power parity exchange rate than the "official" rate, the use of this market rate would reduce our estimate of China's foreign trade dependency ratio by 40 percent. See Dai Gang, "Booming Foreign Ventures in Shanghai," *Beijing Review*, Vol. 30, No. 51, (December 21-27, 1987), p. 28.
34. State Statistical Bureau, *Statistical Yearbook 1986, op. cit.*, p. 499.
35. Based on statistics for imports and exports for total world trade and individual countries in *ibid.*, p. 730 and World Bank, *op. cit.*, pp. 196-197.
36. State Statistical Bureau, *Statistical Yearbook 1986, op. cit.*, p. 92.
37. Estimated from annual absolute levels of employment in each type of ownership and annual total value of retail sales by type of ownership; *ibid.*, pp. 92, 446; State Statistical Bureau, *Zhongguo 1987, op. cit.*, p. 17.
38. Urban collectives include township and village enterprises, while rural collectives include individuals and specialized households.

39. See *supra* note 37.
40. Sales by peasants to non-agricultural residents.
41. Economic Reform Research Institute, Comprehensive Investigation Group, ed., *Gaige: Women Mianlin de Tiaozhan yu Xuanze* (Beijing: Chinese Economics Publishers, 1986). The results of the survey presented here are taken from William Byrd, "The Atrophy of Central Planning in Chinese Industry: The Impact of the Two-Tier Plan/Market System," paper presented to the Arden House Conference on Chinese Economic Reform, Harriman, New York, October 9-12, 1986.
42. These estimates are from Louise do Rosario, "Course Correction," *Far Eastern Economic Review, July 16, 1987, pp. 69-71. A slightly different set of statistics, but following the same trends, was published in a Chinese source: Gao Shangquan, "Progress in Economic Reform (1979-86)," Beijing Review*, July 6, 1987, pp. 20-24. According to Gao, "the number of commodities controlled directly by the Ministry of Commerce has been reduced from 188 in 1979 to 23." Gao presents the same statistics for the decline in the share of investment covered by unilateral budget grants as given in the *Far Eastern Economic Review* article, but adds that "sales of commodities with prices fixed by the state account for only 40 percent of the country's total sales volume."
43. For example, the imports of producers' goods from the Soviet bloc that accounted for 40 percent of investment in machinery and equipment during the 1950s, as well as the 18 thousand Soviet specialists stationed in China, would suggest a very significant transfer of technology took place during that decade. See Robert F. Dernberger, "The Transfer of Technology to China," *Asian Quarterly*, No. 3, 1974, pp. 229-252.
44. At the outset of this section, recognition should be made of the collaborative research project being carried out by Professor Thomas Rawski and his American and Chinese collaborators that promises to contradict many of the arguments presented here. Correctly arguing that the use of official, unadjusted statistics for input and output has resulted in estimates of total factor productivity that seriously underestimate the growth in total factor productivity in Chinese industry, Rawski and his colleagues come up with new estimates for total factor productivity by using adjusted series for capital and labor inputs to remove non-productive inputs, adjusted price indices to deflate the value series to constant prices or real changes in

physical inputs, and deriving weights for the various inputs from econometric estimates for production functions in China's industry. Although the estimates for the pre-1978 period show a higher growth in total factor productivity than the estimates cited here, they do not contradict the fundamental argument being made. As for the post-reform years, Rawski's estimates are considerably higher than the previously available estimates, partly because they rely on the inclusion of data for 1984 and 1985, years of very rapid output growth, not available when the earlier estimates were made. Even so, Rawski's estimates for the increase in total factor productivity in industry during the 1980s would have these increases accounting for two-thirds of total growth in industry, which would contradict both the quantitative data for rapid increases in inputs and the descriptions of industrial developments provided by the Chinese themselves. Thus, I believe that the Rawski estimates may over-compensate for the downward biases in the earlier estimates for total factor productivity growth. As their research continues, Rawski and his colleagues will undoubtedly improve upon their estimates. In the meantime, however, we will rely upon the published estimates presented here, which still show an increase in the contribution of productivity growth to total industrial growth as a result of the reforms; recognizing that the results of research now underway may well determine an even greater contribution due to productivity growth.
45. Robert F. Dernberger, "Financing China's Development: Needs, Sources, and Prospects," *op. cit.*, Appendix Table A-2: China's Economic Growth, 1953-1985; Analysis: Sources of Growth.
46. World Bank, *China: Long-Term Development Issues and Options* (Baltimore: The Johns Hopkins Press, 1985), p. 111 and Table 7.1.
47. Gregory Chow, *The Chinese Economy* (New York: Harper & Row, 1985), pp. 119-131.
48. Gerald M. Meier, *Financing Asian Development: Performance and Prospects*, Agenda Report No. 6 (New York: University Press of America for the Asia Society, 1986), p. 7.
49. For example, see the discussion of the alternative estimates for total factor productivity in industry discussed *supra* note 44. As explained there, part of the reason why the more recent estimates show a much more rapid growth of total factor productivity in the 1980s is due to the addition of statistics for 1984 and 1985, not available when the earlier estimates were

made.
50. Justin Yifu Lin, "Measuring the Impacts of the Household Responsibility System on China's Agricultural Production," unpublished paper, Department of Economics, University of Chicago, April 1986 (mimeo). In a later paper, Lin used a different methodology to estimate the impact of the economic reforms in agriculture and obtained a higher contribution to the increase in output due to the change from cooperative farming to household farming, i.e. he estimates that this change contributed 60 percent of the 26.4 percent increase in output between 1980 and 1983. See Justin Yifu Lin, "Household Farm, Cooperative Farm, and Efficiency: Evidence from Rural De-Collectivization in China," Discussion Paper No. 533 (New Haven: Economic Growth Center, Yale University, 1987).
51. Estimated from the annual value of output in each sector within agriculture in 1983-1986 from State Statistical Bureau, *Zhongguo 1987, op. cit.*, p. 24.
52. *Ibid.*
53. State Statistical Bureau, PRC, *Statistical Yearbook of China, 1985* (Hong Kong: Economic Information Agency, 1985), p. 298; *Statistical Yearbook 1986, op. cit.*, p. 177.
54. State Statistical Bureau, *Zhongguo 1987, op. cit.*, p. 23.
55. Gross value of output, net value of fixed assets, and employment in state-owned industry estimated from annual values of rates of growth in State Statistical Bureau, *Statistical Yearbook 1986, op. cit.*, pp. 21, 92, 224.
56. Gross values of industrial output in 1980 and 1985 from State Statistical Bureau, *Statistical Yearbook 1986, op. cit.*, p. 227, and net value of industrial output in 1980 and 1985 from *Zhongguo 1987, op. cit.*, p. 229. The reason why this result is obtained may be that the gross values are in constant prices, while the net values are in current prices; which means that the use of gross values does bias our estimates of total factor productivity, i.e., the estimates would be lower if we were to use the net value of output *in constant* prices.
57. State Statistical Bureau, *Statistical Yearbook 1986, op. cit.*, p. 47.
58. State Statistical Bureau, *Statistical Yearbook 1985, op. cit.*, p. 375; *ibid.*, p. 271.
59. On the basis of their revision of the available statistics, Rawski and his colleagues would disagree with this statement.
60. State Statistical Bureau, *Statistical Yearbook 1986, op. cit.*, p. 365.

61. Statistics for labor force participation rates and level of unemployment in 1952 and 1985 from *ibid.*, p. 71 (total population), p. 92 (total labor force), and p. 104 ("job-waiting" rate).
62. *Ibid.*, p. 40 (per capita national income) and p. 556 (per capita consumption).
63. *Ibid.*, p. 596. Per capita meat consumption is the sum of per capita consumption of pork, beef and mutton, and poultry.
64. *Ibid.*, pp. 595, 599.
65. Index in current prices *ibid.*, p. 557. Index in constant prices estimated by using growth rates for peasants and non-agricultural residents annual consumption in comparable prices in aforementioned source. Absolute levels in current prices and rates of growth in comparable prices for 1986 from *Zhongguo 1987, op. cit.*, p. 98.
66. At an annual rate of increase of 6.7 percent, compared with 5.5 percent for non-agricultural residents, the peasants' per capita consumption would not become equal to that of the non-agricultural residents until the year 2071.
67. This was often mentioned as the most serious problem in Shanghai at the time of my visit to Shanghai in June 1987.
68. Figures for 1952 in State Statistical Bureau, *Statistical Yearbook 1986, op. cit.*, p. 556. Other years estimated on the basis of indices given in Table 13.
69. The term "falling behind," of course, must have a comparative reference point. Inasmuch as the per capita income of the urban worker remains much higher than that of most peasants, how can the urban workers believe they are falling behind? Simple, they believe their incomes should be a multiple of what the peasants receive, inasmuch as the peasant can rely on self provided consumption and they cannot, i.e. their needs are greater. According to Martin Whyte, a knowledgeable observer of both urban workers and rural peasants, "many urbanites are becoming irate about the 'violation' of the 'natural order of things' because of the extent to which some portion of the peasantry is succeeding in 'getting rich.'" Martin King Whyte, "Social Trends in China: The Triumph of Inequality?" in A. Doak Barnett and Ralph N. Clough, eds., *Modernizing China: Post-Mao Reform and Development* (Boulder, Colorado: Westview Press, 1986), p. 115.
70. An even more striking symbol of the peasants' increasing

incomes and savings I came across during my recent visit to China (Summer 1987) was the various housing and business office projects being constructed in the Western suburbs of Beijing by peasant construction crews financed by peasant savings. These projects were neither small-scale, nor of simple design.

71. This received theory is now under attack due to the experience of more recent successful development efforts, i.e., Taiwan and South Korea. See, for example, J. Fei, G. Ranis, and S. Kuo, *Growth with Equity: The Taiwan Case* (New York: Oxford University Press, 1979).

72. Questions such as this involve people's perceptions and expectations and cannot be illustrated with national statistics for the class-size distribution of income. Each individual makes comparisons with those in his or her reference group and those comparisons are not based on statistics, but on impressions, etc. Thus, statistics such as those presented and analyzed in this paper for the results of the economic reform program may not be all that relevant in determining political support for the economic reform program, i.e., "good" statistical results cannot be directly translated into political support for the reforms as is implied by the statement in the text.

73. In the absence of data to directly estimate the class-size distribution of income in China, Irma Adelman and David Sunding have utilized a statistical interpolation method to utilize the distribution of rural incomes, the distribution of urban incomes, and the differences between the means of peasant and urban incomes to argue that the worsening distribution within the rural sector was more than offset by the reduction in the urban-rural differences so as to reduce the class-size distribution of income for the country as a whole. Not only do their estimates show income inequalities declining after 1978, income inequality in China—according to their estimates—is among the lowest in the world. It should be noted that the data used to make these estimates of the impact of the economic reforms on the distribution of incomes was limited to the years 1978-1983. See Irma Adelman and David Sunding, "Economic Policy and Income Distribution in China," *Journal of Comparative Economics*, Vol. II, No. 3, September, 1987, pp. 444-461.

74. State Statistical Bureau, *Statistical Yearbook 1986, op. cit.*, pp. 576, 582.

75. See *supra* note 71.
76. Rather than use a statistical interpolating technique to utilize what data is available to estimate changes in the class-size distribution of income in China (the approach used by Adelman and Sunding to make up for the lack of the necessary data), Martin Whyte relied upon logic and the marshalling of analytical arguments using rather sound logical arguments in running through the many pros and cons to be evaluated in determining what happened to the distribution of income since 1949. He comes to the conclusion that, although we cannot say "that everything is becoming more equal under Deng," the forces at work are such that we "cannot simply *assume* that everything is becoming more unequal." Martin Whyte, *op. cit.*, p. 120.

FIVE

PROPERTY RIGHTS, ECONOMIC ORGANIZATION AND ECONOMIC MODERNIZATION
DURING THE ECONOMIC REFORMS

Ramon H. Myers, Stanford University

Between 1949 and 1978, the socialist Chinese Communist Party (CCP) radically transformed the human and physical landscape of China, imposing changes so far-reaching as to be unprecedented in that country's history. The replacement of private markets by bureaucratic networks and the nationalization of private property by creating state and cooperative forms of organization were probably the most dramatic, transformative changes that society had ever experienced. The new planned economy, mixed with self-sufficient-type organizations created by the CCP, attracted great interest from outside China. But the regime released little reliable information about how this system performed and operated during a 30-year period. Consequently, outsiders' understanding of how institutional change and these new organizations actually worked has been extremely poor.

Then a dramatic shift in the CCP's line occurred in 1978: to ignore class conflict and concentrate on upgrading the society's "forces of production" in order to modernize the country in the coming decades. New economic reforms followed, aimed at altering the allocation of resources between economic sectors, reforming the price sector, and creating new organizational rules and operations in agriculture. In 1984, the CCP announced that new urban reforms would also take place, such as reorganizing property rights, extending price reforms, and changing organizational rules and operations in state and cooperative enterprises.[1]

In order to understand the significance of these urban economic reforms, we need an analytical framework that can interpret how recent state-induced institutional change, changes in transaction costs, and new organizational forms and behavior are likely to influence economic development in that huge country. The first section outlines such a framework. The second section applies that framework to explain why the CCP decided to initiate new economic reforms, first in 1978 and again in 1984. Section three describes the experimental changes of the past few years as they relate to the urban economic sector. The final section applies the framework to speculate how those reforms might influence future economic development.

AN ANALYTICAL FRAMEWORK

In recent years, a large theoretical literature has evolved focusing on how certain costs—transaction costs—are influenced by institutions, whether of the state or the private sector. The following account briefly describes how institutional change can influence transaction costs, and how those changes are then reflected in organizational forms, behavior, and performance.

Taking our cue from the recent work of Douglass North, we define institutions as a "set of rules, compliance procedures, and moral and ethical behavior norms designed to constrain the behavior of individuals in the intent of maximizing the wealth or utility of principals."[2] The making of these rules, whether by the state or private organizations, is done according to the perceived costs of compliance. To elicit compliance with rules, there are costs for enforcement by those imposing the rules. There are also costs by the parties—agents—of those who carry out transactions within the environment in which those rules operate. If organizations can reduce those costs on a unit basis of output, they can operate more efficiently while choosing the best technology and undertaking innovations. But how can organizations adjust and reduce these costs?

We define these costs as transaction costs. R.H. Coase first identified the existence of such costs in his celebrated 1937 article where he posed the important question of why the firm emerges in the real world.[3] According to Coase, the size of the firm will be determined by how well it manages the costs of using the price mechanism. He defined those costs as transaction costs. If the firm assumed some of those costs by producing the same services which those transaction costs of using the market were designed to obtain, the firm expanded in size and diversified in function. In effect, the evolution of the firm meant that a range of market activity was taken over by the firm and organized through its authority and direction, particularly in the firm's factor market.

A large literature evolved in the 1970s and focused largely

on identifying types of transaction costs and their significance for organizational form and behavior. Alchian and Demsetz, for example, pointed out that firms must monitor the hiring and coordinating of inputs to prevent shirking and ensure quality control over input use under complex conditions of specialization and division of labor.[4]

Effective monitoring also meant transacting with other parties and acquiring correct information, which could be costly. Jensen and Meckling identified the costs of a principle unit transacting with an agent unit as agency costs.[5] George Stigler stressed the high costs involved for any firm or individual to obtain correct information.[6] Then Yoram Barzel cited the measurement costs incurred by firms or individuals to use markets effectively to ensure that the quality of produced or traded items was appropriate for the parties involved.[7]

As transaction costs became better understood, many realized that transaction costs could not be included in conventional accounting concepts or those of microeconomic analysis. Therefore, how did organizations try to reduce transaction costs rather than relying on markets to incur such costs? The empirical work by Alfred D. Chandler, Jr., for example, demonstrated that the innovations by American corporations in the 19th and 20th centuries to reorganize their functions were really innovative means of coping with certain transaction costs.[8] Oliver E. Williamson tried to show that American corporations developed complex, hierarchical and multi-functional operations that substituted for a range of market activity that normally provided those same services.[9] The corporation produced those services more cheaply than relying on the market to purchase them. By doing so, the corporation achieved economies of scale, greater specialization, and more division of labor, thereby increasing its output per unit of input as well as its output per unit of transaction costs. The emergence of internal enterprise hierarchy represented a partial substitution for the market system, but a type of substitution that made more productive use of resources.

Economic historians like Douglass North then observed that a link possibly existed between institutional change, transaction costs, organizational forms and economic development. This line of thinking argues that changes in transaction costs and/or production costs for organizations imposed their economies of scale and productivity. But institutional change, whether initiated by the state or taking place in society, influences the structure of incentives for individuals and their organizations to cope with transaction costs and/or production costs more effectively. If new incentives enable agents and their organizations to contract with each other to enhance their specialization and division of labor, productivity and output can be increased, so that additional gains accrue to owners of resources. For example, Douglass North and Robert Paul Thomas argued that when common property rights over arable land became privatized some 10,000 years ago, the new resource owners were able to reap the gains of specialization and trade in agriculture, making possible the first economic revolution in mankind's history, whereby communities centered around agriculture and commerce began to develop.[10]

North and Thomas also argued that when institutional change in the factor markets occurred during the Middle Ages, such as the shift from lord-serf to landowner-tenant contract relations, the incentives for resource owners to use the market greatly improved.[11] They also noted that the improvements in military technology in the early modern period, when initiated by various monarchies, greatly reduced brigandage and piracy, so that commerce and manufacturing could flourish.[12] These institutional reforms and technological improvements were closely linked. When institutional change reduced transaction costs, agents and their organizations also benefited from technological change.

Later improvements in commercial law and clarification of property rights also made possible the rise of trading companies and the corporation, whereby agents and these organizations developed new economies of scale and more efficient production and distribution.[13] Alfred D. Chandler, Jr.'s account of the late 19th-century American railway corporations described the "decentralized line-and-staff concept of organization" that gave on-line managers the authority and responsibility for the men involved with the basic function of the enterprise; other functional managers (staff executives) actually set new standards.[14] This new organizational compartmentalization made possible greater geographical division and functional efficiency for building and operating railroads. Other corporations evolved later with these same new hierarchical structures to achieve a forward integration into manufacturing as well as with distribution.[15]

Similarly, transaction costs analysis has been applied to socialist-planned economies to determine why economic growth slowdown and poor technological innovation characterized the Soviet Union since the 1960s and 1970s. John H. Moore examined the role played by bureaucratic organs in the Soviet Union and concluded that high agency costs characterized their activities, so that those bureaucratic hierarchies greatly impeded enterprise capabilities to introduce technological change.[16] In this case, state bureaucratic hierarchy had substituted for the range of market functions that enterprises could have derived advantages, and the lack of economies of scale inhibited innovative changes. The dysfunctions of this bureaucratic hierarchical system flow from the system's high agency costs to monitor enterprise performance and activity to ensure that managers complied with the plan and state rules.

PROPERTY RIGHTS, ECONOMIC ORGANIZATION AND ECONOMIC MODERNIZATION

While the linkage between institutional change, transactions costs, organizational form and behavior, and economic development is unclear in history, some linkage patterns seem to unquestionably exist. For example, a country like the United States certainly experienced profound sectoral change and rapid economic transformation since the early 19th century. At the same time, the organizations handling transaction costs underwent complex change, as Chandler and others have noted. But the total transaction costs per unit of output seem to have declined over this same period, as Wallis and North have recently found.[17] Even though transaction costs increased, production costs in the American economy probably substantially declined to offset this increase.[18] The fall in production costs had been made possible by greatly increased specialization and division of labor. But those two developments only seem to have been made possible by the new organizational forms Chandler and others have observed. Those new organizational forms were associated with higher transaction costs. But the huge increase in productivity and output that also occurred then made possible the decline in total transaction costs per unit of output.

Keeping in mind that total transaction costs consist of measurement costs (to include information, monitoring and contracting costs) and enforcement costs, we can show how changes in transaction costs are linked with economic development in different economic systems.

Successful economic development means reducing the total costs of production. The total costs of production comprise transaction costs and production costs. In economic systems with private markets and institutions that favor private transactions and private property, the total costs of production appear to be reduced over time in the following way.

If institutional change encourages more transactions through the market and/or for organizations to internalize some of their transaction costs, total transaction costs might rise. But at the same time, the new types of transactions undertaken by organizations appear to become associated with greater economies of scale and new technological change. In other words, more organizations are able to purchase new capital, select the best technology and increase their productivity. The rapid fall in production costs flowing from these developments could still be associated with higher total transaction costs (or declining transaction costs). Whichever the case, if the economic system has experienced sustained increases in productivity along with a higher growth rate of output for goods and services, the transaction costs per unit of output still might fall.

Any economic system experiencing these broad developments has probably incurred higher economic growth rates, rising factor productivity, improved income distribution and greater consumer welfare. The economic history of such an economic system would also reflect complex organizational developments that appeared to facilitate even more technological change and innovation as time passed. Market competition would prevent monopoly and oligopoly organizations from capturing the lion's share of the gains from specialization or receiving high economic rents. The complex process just described seems to have reflected the successful economic development of the United States, other Western countries, Japan and certain Pacific Basin nations since the Industrial Revolution.

What of the development pattern of socialist systems? Institutional change redistributes property by state nationalization of wealth and the creation of state and collective organizations. When such a state opts for a planned economy, a bureaucratic hierarchy substitutes for the marketplace. State organs monitor the activity of all organizations. The elimination of both product and factor markets and their replacement by a bureaucratic hierarchy to determine those resource allocations will be associated with higher total transaction costs. Measurement costs related to the difficulties of quality control over output greatly increase, as do agency costs (monitoring costs). These total transaction costs increasé for the enterprises, as well as for state regulatory agencies.

This new bureaucratic hierarchy must create an artificial price system. But these new prices no longer represent consumer demand, and they will not provide the correct information for organizations to coordinate their inputs and transform them into output of higher value. Enterprises only have incentives to maximize their planned output value irrespective of how they use their inputs. Enterprises are not under any pressure to maintain quality control over their production. Because managers and their workers are rewarded according to fulfilling their plan, no incentives exist for them to improve quality of production, accumulate new capital or select new technology unless these are strictly related to that firm's plan. In most instances, they are not. Agency costs outside and within the firm remain high to ensure that all organizations comply with planners' rule and preferences. Organizations try to comply with bureaucracy preferences because no formal markets exist to allow them any alternative. Bureaucrats willingly subsidize enterprises to meet their plan even when they perform poorly and produce low-quality goods and services. The artificial price system used by enterprises generates losses for some firms and gains for others. Yet bureaucrats provide subsidies to failing firms to offset their losses. No bureaucrats are held accountable, because their bureaucratic organs can allocate those losses to the entire bureaucratic hierarchy's budgeting and not to any specific bureaucratic unit.

Because organizations appear to have little incentive to

improve the productivity of their capital and labor, few achieve economies of scale. Specialization and trade remain limited to a few enterprises favored by certain bureaucratic organs. Therefore, production costs do not decline rapidly, if at all. Because total transaction costs in the economy have risen considerably, the negligible or nonexistent decrease in production costs will mean that total costs in this type of economic system remain high over time or decrease only very slowly.

To make matters worse, this kind of total cost performance seems to be associated with little or no productivity growth. The poor-quality output of goods and services also brings minimal gains in consumer welfare. The economy still might achieve a high rate of economic growth, but declining factor productivity and high total costs will eventually cause a slowdown in the economy's total growth rate. The initial redistribution of property rights and wealth produces a much more equal distribution of income than previously. But the general economic performance of this economic system is characterized by a great waste of resources, minimal gains in consumer welfare, and low productivity.

For socialist-planned economies, the above process of linking total costs, made up of transactions costs and production costs, with economic development differs greatly from the process of development described for capitalist systems. But it tells us a great deal about how the Chinese economy performed after 1949.

CHINA'S ECONOMIC DEVELOPMENT: 1949-1978

By 1957, state institutional reforms had nationalized all property and eliminated the market system and replaced it with a bureaucratic hierarchy to regulate all transactions between organizations. We know little about how this new bureaucratic hierarchy of planned economy really worked between 1957 and 1978. Generally speaking, we can say that national ministries took charge of all economic sector activities. Provincial and municipal bureaucratic organs assisted these national ministries. Networks of bureaucracies regulated how firms acquired and used raw materials and capital goods, processed commodities, and produced final commodities as well as supplied funds, labor, and technology. Crisscrossing networks of bureaucratic hierarchies handled all transactions for state and cooperative organizations to acquire their inputs and to sell their outputs and services. This new system relied upon an artificial price system as a means for accounting and making valuations.[19]

Information from informants in Hong Kong and accounts that surfaced after 1978 confirmed that this new bureaucratic hierarchical system had to be supplemented by an "informal" system of transactions (black markets), only tolerated by the authorities because it allowed the "formal" system to work. This informal system also relied upon personal connections and "bureaucratic favors" to handle transactions that the formal system failed to accommodate. Whether acquiring scarce inputs or forging performance reports, enterprises had to rely on this informal system in order for the formal bureaucratic hierarchy to perform.

The number of state-owned firms in industry alone reached 59,600 in 1957, declined to 45,000 in 1965, but expanded to 83,700 in 1978.[20] Similarly, collective-owned enterprises between 1957 and 1965 remained fairly constant at around 110,000 but more than doubled to reach 264,000 in 1978.[21] The expansion of these state- and collective-owned enterprises in industry corresponded to the disappearance of privately owned industrial enterprises.

This new system soon became riddled with scarcity, poor quality products and services, inefficiency and waste, even though the capital stock greatly expanded. For state- and cooperative-owned enterprises, capital stock more than doubled every decade at an annual growth rate well above 12 percent.[22] But the profit earned per unit of capital (valued at a 100 *yuan*) declined over much of the period.[23] Industrial output value per unit of fixed capital (valued at 100 *yuan*) also declined after 1957, as did that for circulating capital as well.[24] For many of the enterprises created in the 1950s through Soviet economic aid, their capital had never been replaced and improved.[25] In the late 1970s, conference reports on the price system described the great enterprise waste and inefficiency produced by the artificial price system over these decades.[26] Sectoral surveys undertaken at the same time alluded to many industries where a large number of firms contributed only a small share of the industrial value compared to that originating from a few dominant firms.[27]

In addition to these dysfunctions of urban industry, many Chinese economists also admitted that this new industrial structure failed to produce sufficient consumer goods and services.[28] They complained that the industrial system created before 1978 rapidly increased industrial output, but much of it could not be used by the Chinese economy.[29] Thus, the structure of the economy and its poor performance became the target of criticism by the late 1970s.[30] The general consensus reached in the party as well as among state officials was that China needed economic reforms to repair the damage. The Chinese economy had to be made more efficient and produce a different mix of output to improve the welfare of the people.

Why had the system performed so poorly, so as to become the target of criticism and reform by the CCP in the late 1970s? The rapid substitution of the market by a bureaucratic hierarchy to regulate enterprise transactions and behavior only worked if planners and officials introduced an artificial price system. The monitoring of managers and workers required the careful checking and regulation of all enterprise decisions and behavior. Enterprises expended a great deal of time and energy to operate. They had no incentive to improve quality control. As their transaction costs increased, these enterprises were discouraged from developing new economies of scale except for a few firms permitted to do so according to the annual industrial plan. The new incentive structures elicited enterprise behavior to maximize planned output value irrespective of output quality and how efficiently inputs were used.

High transaction costs were associated with an organizational structure within enterprises that made it very difficult to achieve economies of scale for facilitating innovation and the efficient use of capital stock. The regulations imposed to govern enterprise transactions prevented mergers, bankruptcies, and transfers of assets between firms. Therefore, enterprise productivity remained low or even declined. The artificial price system and use of subsidies rewarded the inefficient firms rather than allowing their exit from the system. As a consequence, production costs failed to decline significantly. Total costs, then, remained very high, and it is very likely that transaction costs per unit of output increased.

INSTITUTIONAL CHANGE SINCE 1984

The recent urban economic reforms represent new institutional changes which the state has promoted or hopes to make. These institutional changes mean that the old decision-making rules for enterprises have been altered or will be changed. These rule changes have not necessarily been widely encouraged across the country; many are at the experimental stage. These new decision-making rules permit organizations to engage in more transactions than the bureaucratic hierarchical system formerly allowed. Five types of rule changes are briefly described below: allowing bankruptcy; transferring assets outside the bureaucratic hierarchy; allowing joint stock enterprises; allowing enterprises to transact with certain designated enterprises; and allowing more decision power to enterprises.

1. A New Bankruptcy Law

By late 1986, debate had focused on how to solve the problem for those state-owned firms which had loaned funds to other enterprises and never been repaid. For example, in November 1986, a factory producing explosives in Shenyang city had incurred losses for 10 years. This state-owned enterprise owed 120 creditors, but had assets of only 220,000 *yuan*. Such cases proliferated. Many managers pleaded for a bankruptcy law.

On November 29, 1986, Peng Zhen, chairman of the National People's Congress (NPC), led the discussion of a new bankruptcy law for state-owned enterprises.[31] Peng favored experimenting with new rules governing bankruptcy in certain parts of the country before introducing a comprehensive law for the country. At the Sixth NPC Standing Committee meeting on December 2, 1986, a trial Enterprise Bankruptcy Law was introduced.[32] This new law, comprising 43 articles in six sections, gives jurisdiction over bankruptcy cases to the People's Court system. The court appoints a committee to handle the bankrupt firm after certain procedures have taken place. Creditors first take their claims to the court. The court then assembles all parties to explore how previous contracts that were not followed can be enforced. If enforcement of old contracts fails, the court then initiates steps for the debtor enterprise to consolidate their assets. A committee ranks all creditors according to priority of claim for compensation. If necessary, the committee liquidates all assets and dissolves the enterprise. Time-consuming as the above steps are, the state seems serious about introducing this new enterprise law.

The enterprise law calls for the State Council to grant enterprise managers greater authority over their enterprises.[33] But passing this law has been delayed, as of 1987. There appears to be strong public support for it. In mid-1986, a survey of public opinion in four cities found that most people were willing to have a bankruptcy law applicable to state-owned enterprises.[34]

2. Transfer of State-Owned Assets to Other Enterprises

The sale of state-owned assets to other enterprises, whether by enterprises engaged in manufacturing or services, gradually occurred in 1986 and 1987. First in Shenyang, then in Beijing and other metropolitan areas, local officials began permitting state-owned enterprises to auction their assets to other buyers. For example, December 1986, a group of state-owned shops were auctioned to private individuals.[35] A 32-year-old state employee of the Haidan Grocery Company outbid other competitors with a 90,000 *yuan* (US $24,000) offer to buy the Lucky Gate Grocery, a 100 square meter shop with a yard on the fringe of the city. At this same auction, a restaurant, a bicycle repair shop and a barber shop went to bidders at offer prices ranging from 70,000 *yuan* (US $19,000) to 110,000 *yuan* (US $30,000). The monies collected from the auction passed into a fund for early retirement of the employees of the formerly state-owned shops or assisting them to find new jobs. These shops had been running at a loss for several years. Officials expressed satisfaction with the successful auction, but said they "would have to evaluate the results of their experiment before deciding whether to put other enterprises on the block."[36]

As of October 1985, there were reported to be 85,805 state-owned retail commercial and catering service enterprises in China. But of that number, 49,085 had been contracted to collective-owned management, 7,843 had been turned into collective-owned enterprises, and 7,377 had been rented to individuals.[37] By leasing and transferring state-owned enterprises in services to collective- and individual-owned enterprises, the state has promoted a flood of small stalls or shops in Beijing and other cities that sell everything from watermelons to sunglasses, repair shoes and watches, roast sweet potatoes over fires in steel drums, or serve lamb kebabs.[38] Because of the great expansion of service enterprises, a city like Beijing is now reported to have more than a quarter-million migrants from the countryside who are employed by or operate these service establishments. While these transients must report to the police and the Public Security Bureau within three days of their arrival in a city, they can receive permission to remain for six months if they are employed, and many remain longer. The police make random checks of urban neighborhoods to look for people without proper household registrations. But the rapid growth of this new urban migrant population appears to be defeating their efforts to force those people to return to their home provinces.[39]

3. Creation of Joint Stock Enterprises

By 1986, experiments were taking place in various cities to allow state-owned enterprises to become joint stock enterprises owned by various types of shareholders. This new decision-making rule change called for creating more collective- and individual-owned enterprises in services and industry. In 1985, Professor Li Yining of Beijing University's Economics Department and other economists argued that reforming the ownership system was the key to successfully reforming the socialist economic structure. Li suggested three different ownership forms for industry: (1) ownership and operation by the whole people (state-owned and -run enterprises); (2) ownership by the whole people and autonomous operation by the enterprise; and (3) renting or contracting by small-scale enterprises.⁴⁰ Li went on to say that large-scale enterprises should remain under state ownership and control, but medium-sized enterprises should become joint stock companies with the state holding the controlling interest and the enterprise and individuals holding the remainder.

He then advanced three principles governing the sale of shares for medium-sized enterprises.⁴¹ First, new enterprises should raise capital by issuing shares. Second, existing firms should also issue shares. Finally, the ratio of shareholdings between the enterprise and the state should be decided by the proportion of state and enterprise capital used for reinvestment. Enterprises should be able to sell their shares to other individuals as well as to their own employees. A board of directors should be appointed to direct the enterprise's activities. It is not clear, however, what kind of stock market has been agreed upon to handle these activities. Nor have the laws to regulate such a market been developed for the country.

But some enterprises already have become joint stock companies. In Wuhan city, for example, the Wuhan Electrical Wire Company had a registered capital of 12 million *yuan* (US $3 million) and 3,000 employees. It set up its first stock shareholder company with 12 factories in and outside of Hupei province.⁴² The state owns half of these shares, each valued at 100 *yuan*, and the rest are owned by the participating factories, cooperating businesses, and individual buyers. The company issued five-year shares worth 410,000 *yuan* and planned to issue three-year shares valued at 3 million *yuan*. Shareholders would share profits and losses, and probably receive 10 percent interest. The company planned to use the funds it had mobilized for purchasing new capital to increase productivity. Similar examples have been reported in cities like Shenyang, Chongqing, Shanghai and Guangzhou. Moreover, people are urging that collective-owned enterprises become joint stock enterprises.[43]

But no reports so far have clarified how sellers learned about the sale of enterprise stock, how the purchases of stock were arranged, and how stockholders would be protected under the law if those enterprises fail to pay the expected dividends. In fact, there has been no national law passed by the State Council to govern the issue of enterprise stock and how a stock market would be regulated. Moreover, since the student movement in December 1986, violent attacks on Western liberalization appeared in the press in early 1987, even criticizing economic reforms like creating joint stock enterprises.

4. Giving Special Rights to State-Owned Enterprises

Just as certain households in the countryside were designated to have special property rights to organize the production of certain commodities—a key feature of the rural responsibility system—so has the State Council recently passed a law arrogating special rights to large-scale state-owned enterprises.[44] The state will allow certain enterprises to have unusual independent decision-making authority but still intends to guide them in their economic tasks. This favorable status requires some special conditions: Those enterprises must be large and strategically located in a given branch of industry, their commodities must generate significant income and/or foreign exchange, and they must be technologically advanced and well managed. Such firms can apply to the State Planning Commission for independent decision-making authority. If the commission approves their applications, they will then be granted a special status and the relevant ministries and departments notified. These special enterprises receive relevant economic targets and additional information to grant them a wide range of independent activity outside the plan, such as participating in special capital construction projects, etc. If the enterprise wants to engage in technical innovation, "it shall not need approval for single-item projects if its overall plan has already been approved by the State Planning Commission."[45] Yet the enterprise still would be required to obtain its materials mainly from the market under state guidance plans and sell in approved markets. A close reading of these new regulations indicates that these special enterprises will have more independence to make innovations, undertake capital investments, and develop multi-product lines for sale abroad or in the domestic market after they have met the targets assigned under the mandatory plan.

5. Rules Allowing Enterprises New Responsibilities

In May 1984, the state initiated a trial plan to permit state-owned industrial enterprises to adopt the plant director responsibility system. By January 1987, some 30,000 enterprises out of more than 400,000 had adopted this system.[46] In fact, the key features of this new system had provided the basis for the new enterprise law that the NPC approved in November 1986 (but which had yet to become law in 1987). This new industrial and commercial responsibility system makes the plant director, rather than the CCP plant secretary, the legal representative of an enterprise, thereby giving the director the authority and power to guide the enterprise.[47] The director appoints management personnel and is responsible to the relevant state agencies. The director's income can be two to four times that of ordinary workers, but the managerial responsibilities are greatly increased. Enterprises also must elect a management committee to assist the plant director in making policy decisions. Such a committee comprises representatives of the CCP and the trade unions, as well as the chief engineer and accountant. Backers of the new enterprise law hope that some 400,000 state-owned enterprises will eventually be managed by bona fide managers, rather than by party cadres. But to achieve this will be difficult, because in many enterprises "these cadres constitute about 30 percent of the total number of workers and staff members."[48]

How has the new system actually worked in different provinces? Sichuan Province initiated this system in 1984 in Chongqing city. The assets of these new enterprises belong to the state, but their directors are appointed by government agencies after being selected on a competitive basis. These directors are responsible for fulfilling the plans drawn up by the state, but they have greater decision-making authority than under the former system, when plant directors were not permitted to manage the enterprise's assets. For example, one plant manager, named Xu Jingcheng, age 47, now appoints his executives, directs production, sets prices and distributes bonuses.[49] Some 19 state-owned Chongqing enterprises have advertised for directors and selected them (with city government approval) on a competitive basis.

Plant directors in Chongqing now use enterprise assets more creatively, establish contracts with other enterprises (subject to city government approval), and hire and dismiss labor.[50] The salaries of directors who have achieved their targets are increased. When enterprise assets rise, their pay also rises. Similarly, if enterprise profits and assets fall, the director's salary is reduced, and he/she can be sacked if losses exceed 10 percent. But the mechanism relating a director's salary and performance is still not clear.

In other cities, like Shenyang in the northeast, this new system has allegedly permitted state-owned enterprises to develop lateral markets and diversify their production. For example, the Shenyang Electric Cable Plant reportedly had 10,000 workers and 300 million *yuan* worth of output. In

1986, it established 14 new sub-factories with managers to handle its business there.[51] After meeting its assigned targets, the enterprise markets and produces what its management decides. The Shenyang Heavy-Duty Machine Building Plant employed 22,000 workers and had 240 million *yuan* in fixed assets. It produced mining, forging and steel-pressing and -rolling equipment. Under this new system, its management diversified production; its 30 new plant projects work through five new distribution channels to market their output.

Returning again to Chongqing city, such enterprises have also been allowed to develop market links outside of the mandatory plan. For example, the Tian Fu Cola Beverage Company now has branches in 23 provinces and cities to market its products.[52] Similarly, in Xiang Fan city of Hubei Province the factory director responsibility system made progress in 1986, although there were a number of reports of newly-appointed factory directors voting themselves large and frequent salary increases.[53]

Perhaps one of the most novel forms of this new enterprise responsibility system is that of city governments raising funds to create new enterprises under collective or individual ownership. Apparently with or without permission from Beijing or the province, some city governments have taken the initiative to create new industrial and service enterprises outside the mandatory planning system. For example, in 1984, Anqing city of western Anhui Province created 596 collective-owned enterprises to produce industrial products.[54] Their commodity sales in 1985 accounted for nearly half of the city's total annual production sales. It was also reported that a service enterprise formed in May 1983, had already built five plants, 17 department stores, two joint ventures and 14 branches and offices in other parts of the country, including Beijing, Tianjin, Guangzhou's Shenzhen Special Economic Zone and Hong Kong.[55] The enterprises in

Anqing enjoyed greater decision-making power, were responsible for their profits and losses, could select their factory directors, and freely set wages. We simply do not know how many examples of this kind have already occurred throughout China.

THE LINK BETWEEN INSTITUTIONAL CHANGE AND ECONOMIC GROWTH

Having briefly outlined some of the decision-making rule changes recently allowed or still pending, are such changes likely to reduce total costs and improve quality control so as to improve China's economic development performance?

On first glance, the above decision-making rule changes would appear to allow state- and cooperative-owned enterprises to engage in more transactions outside the bureaucratic hierarchical system. These rule changes also probably facilitate those "informal" transactions that long complemented the bureaucratic hierarchical system. As Dorothy Sollinger has found from her field work in Wuhan city, enterprises long used "informal" transactions and since 1984 the reforms to give Wuhan greater autonomy have merely legitimated those transactions.[56]

If enterprises can transact more freely with each other to exchange assets, purchase inputs, sell their output, and introduce innovations, then transaction costs rise. Will these new transactions improve quality control to help reduce production costs? And will these new transactions permit more specialization and trade and allow all enterprises to capture the income gains, so that they can achieve greater economies of scale and higher efficiency? The slight improvement in total factor productivity for manufacturing since 1978 suggests that perhaps the economy has benefited from these reforms.

But CCP and state officials also worry about the unintended consequences likely to flow from these new reforms. They fear that if enterprises have greater independence to transact with each other, those activities will divert resources away from the mandatory planned sector, increase inflationary pressures, produce greater economic instability, promote increased income distribution inequality, and encourage more illegal activity (corruption, etc.). In order to prevent these new developments, the authorities are creating new bureaucratic organs and decision-making rules to supervise and regulate those enterprises transacting with each other outside the mandatory planned system. The new web of rules and official organs that will monitor these enterprise transactions is only now taking form.

We have a few examples of these new trends from the recent field work of an American expert. Dorothy Solinger describes recent efforts by Wuhan authorities to obtain permission from provincial and Beijing officials to allow its enterprises to transact with other enterprises outside of Hubei Province. While some progress along these lines was achieved by Wuhan's officials, "the State Planning Commission wants to put some 500 of the largest enterprise groups under its own control, and a lot of the unions of all sizes are suffering from administrative interference."[57] Even within Wuhan, officials have set up two new committees to approve and supervise new transactions allowed by Wuhan enterprises. One is an Economic Levers Adjustment Committee and the second is the Wuhan Economic and Technology Cooperation Committee. The former is empowered to give approval for prices, taxes, and bank loans for enterprises, whereas the latter is designed to promote lateral exchanges and help solve their problems. But Sollinger concludes that the establishment of these two new organs "is a state-sponsored effort to keep a grip on the growing trade."[58]

Just as every new reform seems to be accompanied by additional bureaucratic regulations to monitor the new transactions enterprises are permitted to make, so are the new reforms also promoting more bureaucratic rivalry and infighting. When the bureaucratic hierarchy permits more enterprise transactions between enterprises, various organs and their officials in the hierarchy sometimes become threatened by these new transactions and risk the loss of their authority, power and privilege. Roy Grow's account of how several enterprises in northeast China tried to engage in new transactions shows that bureaucratic rivalry soon followed.[59]

The China No. 1 Stamping Plant nearly managed to arrange a contract with the Dalien Canning Plant to supply canning jar lids. Such a contract would have allowed the canning plant to increase production and sales of canned vegetables to an American firm that would have offered top prices, thus enormously increasing the plant's sales revenues. The China No. 1 plant would have gained additional sales revenue to purchase from a Japanese company the new production equipment necessary to upgrade its production and efficiency. The contract collapsed, however, because of opposition from Shenyang administration officials who refused to approve it. Their refusal was predicated on the information that several provincial branches had been courting a particular Japanese trading company to consider investing in several regional projects, but that Japanese trading company refused to agree to that deal because it had learned that the American company purchasing vegetables and grain from Dalien had been given exclusive purchasing rights in that region, thereby excluding its operations. Although the Dalien regional officials had approved the contract between China No. 1 and the Dalien canning factory, the Shenyang regional officials disapproved. The bitter contest between the two regional official organs had to be resolved by officials in Beijing. According to Grow, even Beijing's officials were not able to sort out the consequences of these new contracts for the regional economy, and failed to resolve the mess. Grow's narrative shows how official turf wars only intensify and proliferate after more enterprises begin to transact with each other.

A third consequence of these urban reforms allowing more transactions between enterprises is to encourage them to use those "informal" channels long available in previous decades to facilitate their transactions in ways never considered by the parties themselves. The flood of complaints from officials around the country about so-called corruption shows that this activity has been closely connected with more enterprise transactions.[60] With more enterprise transactions taking place, new opportunities become available for managers and their workers to use their new contracts for including "personal" transactions unrelated to the activities of the enterprise. Yet these additional, "personal" transactions often facilitate the enterprise transactions themselves. Even so, many officials perceive these new transactions very differently and refer to them as corrupt behavior dangerous to the socialist ethics of Chinese society. To respond to this problem, officials have tried to impose new controls to monitor such behavior. Naturally, more official regulations become associated with more private transactions.

The above three consequences from the recent reforms are bound to increase transaction costs for enterprises. These new enterprise transactions will increase sales, improve economies of scale, and upgrade efficiency. But will these improvements be sustained and significant enough to reduce production costs over time? Furthermore, will quality control over output be increased substantially to lower production costs? Or will these reforms be associated with more controls, which will increase transaction costs even more? Furthermore, will bureaucratic conflicts provide enough negative incentives to discourage enterprises from engaging in more transactions with each other? More time will have to elapse before these questions can be answered. In the final analysis, if production costs are not reduced substantially as more transactions take place between agents and organizations, total costs will only fall slowly or not at all. The complex interaction between transaction and production costs will be crucial to determine which direction total costs will go.

NOTES

1. See "Decision of the Central Committee of the Communist Party of China on Reform of the Economic Structure," *Beijing Review* 27:44, October 29, 1984, pp. iii-xvi.
2. Douglass C. North, *Structure and Change in Economic History* (New York and London: W.W. Norton & Co., 1981), pp. 201-202.
3. R.H. Coase, "The Nature of the Firm," *Economica* n.s. 4:16, 1937, pp. 386-405.
4. Armen A. Alchian and Harold Demsetz, "Production, Information Costs and Economic Organization," *American Economic Review* 62:5, December 1972, pp. 777-795.
5. Michael C. Jensen and William H. Meckling, "Theory of the Firm: Managerial Behavior, Agency Costs and Ownership Structure," *Journal of Financial Economics* 3:4, October 1976, pp. 305-360.
6. George J. Stigler, "The Economics of Information," *Journal of Political Economy* 69:3, June 1961, pp. 213-225.
7. Yoram Barzel, "Measurement Cost and the Organization of Markets," *Journal of Law & Economics* April 24, 1982, pp. 27-48.
8. Alfred D. Chandler, Jr., *The Visible Hand: The Managerial Revolution in American Business* (Cambridge, Massachusetts, and London: Belknap Press of Harvard University Press, 1977), in particular, see Chapters 5, 6, 9 and 14.
9. Oliver E. Williamson, "The Modern Corporation: Origins, Evolution, Attributes," *Journal of Economic Literature* 19, December 1981, pp. 1537-1568.
10. Douglass C. North and Robert Paul Thomas, "The First Economic Revolution," *Economic History Review* series 2, 30:2, May 1977, pp. 229-241.
11. Douglass C. North and Robert Paul Thomas, *The Rise of the Western World: A New Economic History* (Cambridge, England: Cambridge University Press, 1973), p. 92.
12. *Ibid.*, pp. 93-101.
13. *Ibid.*, pp. 146-156.
14. Quoted in Williamson, "The Modern Corporation," *op. cit.*, pp. 1553-1554.
15. *Ibid.*, p. 1552.
16. John H. Moore, "Agency Costs, Technological Change, and Soviet Central Planning," *Journal of Law and Economics* 24,

October 1981, pp. 189-214. Moore ignores quality control in central planning systems. Because market competition is nonexistent in such systems, enterprises have no incentive to maintain quality when contracting with other agents to supply products or services, because the "plan" determines the terms of contracting.
17. See John Joseph Wallis and Douglass C. North, "Measuring the Transaction Sector in the American Economy, 1870-1970," in Stanley L. Engermann and Robert E. Gallman, *Long-Term Factors in American Economic Growth* (Chicago: University of Chicago Press, 1986), p. 121.
18. I owe this observation to conversations with Douglass North.
19. The discussion here follows some of the analysis developed by North, *Structure and Change in Economic History, op. cit.*, Chapter 15.
20. *Zhongguo jingji nienjian* [Almanac of China's Economy] (Beijing: Beijing jingji guanxi qubanshe, 1986), pp. iii-18.
21. *Ibid.*
22. For state-owned enterprises, *ibid.*, pp. iii-26; for collective-owned enterprises, *ibid.*, pp. iii-27.
23. *Ibid.*
24. *Ibid.*
25. *Renmin ribao*, November 30, 1986, p. 1.
26. *Shehui jhuyi jingji zhong jihua yu shichang di guanxi* [The Relationship Between Plan and Market in Our Socialist Economy] (Beijing: Zhongguo shehui keshue qubanshe, 1980), I: pp. 40-41; II: pp. 730.
27. Ma Hong and Sun Shangqing, eds., *Zhongguo jingji jieguo wenti yanjiu* [Studies of the Problems Concerning China's Economic Structure] (Beijing: Renmin dachangshe, 1981), Vol. 1-2.
28. Such discussion can be found in George C. Wang, ed. and translator, *Economic Reform in the PRC: In Which China's Economists Make Known What Went Wrong, Why, and What Should Be Done About It* (Boulder, Colorado: Westview Press, 1982), Chapters 4 and 6.
29. *Ibid.*, Chapter 4.
30. *Ibid.*, Chapters 5 and 10. Also Xu Dixin *et al.*, *China's Search for Economic Growth: The Chinese Economy Since 1949* (Beijing: New World Press, 1982).
31. US Foreign Broadcast Information Service, *China Report: Economic Affairs*, January 16, 1987, p. 34. But see Li Peng's

response to newsmen's questions at a March 28, 1987, press meeting, in *Daily Report: China*, Vol. 87-060, March 30, 1987, K3: "When conditions are right, these rules and regulations [the enterprise law] will be embodied in laws." At the same meeting, Tian Jiyun said, "The Bankruptcy Law has already been officially approved by the NPC, and the implementation of the new law starts this year. Of course, the comprehensive implementation of the Bankruptcy Law will begin after the adoption of the State Enterprise Law by the NPC. At present, we are conducting experiments for the implementation of that law."

32. "Text of Enterprise Bankruptcy Law," *Daily Report: China*, Vol. 86-234, December 5, 1986, pp. K1-K7.
33. *Daily Report: China*, Vol. 87-060, March 30, 1987) p. K3; see also *Daily Report: China*, Vol. 87-066, April 7, 1987. p. K27.
34. "Poll on Public Acceptance of Bankruptcy Law," *China Report: Economic Affairs*, December 1, 1986, pp. 7-10. Polls showed that "the idea of 'feeding off the big rice pot' has gradually faded from public consciousness and people are able to adapt to competition and risk-taking, up to a point" (p. 9).
35. For Beijing, see "Beijing Holds First Auction of State-Owned Shops," *China Report: Economic Affairs*, December 15, 1986, pp. 23-25. For Shenyang's auction of a state-owned explosives factory, see *Renmin ribao*, August 4, 1986, p. 1.
36. "Beijing Holds First Auction of State-Owned Shops," *loc. cit.*
37. Guo Dongle, "Is it Retrogression to Contract Small State-Owned Retail Shops to the Collective or Rent them to Individuals for Their Management?", *China Report: Economic Affairs*, March 27, 1986, pp. 79-81. See also the argument for state enterprises to lease assets from foreign enterprises, in "Why Is Leasing Regarded as an Effective Way to Import Technology and Equipment?" ,*China Report: Economic Affairs*, October 29, 1986, pp. 76-78.
38. Edward A. Gargen, "China's 'Wave of the Future': Moving to the City," *The New York Times*, May 4, 1987, p. 10. For other examples of the state-owned enterprises leasing shops, etc., see Gao Dichen, "A New Attempt to Reform State Commerce Operations: An Investigation of Integrating Leasing, Contracting, and Share Capital Operations," *Daily Report: China*, Vol. 87-081, April 28, 1987, pp. K16-K21.
39. Gargen, *loc. cit.*
40. Chao Yu-sun, "Changes of Shareholding and Ownership in

Mainland Enterprises," *Issues & Studies* 23:2, February, 1987, p. 10. But for general reviews and debates over enterprise shareholding, see Gao Qinglin, "China's Shareholding System is Advancing Amid Controversies," *China Report: Economic Affairs*, February 13, 1987, pp. 1-6; and Liu Guangdi, "We Must Not Exaggerate the Role of the Shareholding System in Reform," *China Report: Economic Affairs*, February 13, 1987, pp. 7-11. Also *Renmin ribao*, August 10, 1986, p. 1;, August 18, 1986, p. 5.
41. Chao Yu-sun, *op. cit.*, p. 11. But other analysts pointed out that the state should not allow joint stock enterprises to form without having a stock market to exchange stock. See Liu Rongcang, "Guanyu toufen jingji di jige wenti" [Some Problems Concerning a Joint Stock Company], *Jingji wenti* 9, 1986, p. 13.
42. "First Shareholder-Owned Enterprises in Wuhan," *China Report: Economic Affairs*, January 16, 1987, pp. 51-52.
43. Zhou Huichun, "An Attempt at the Joint Stock System in Collective Enterprises," *China Report: Economic Affairs*, January 23, 1987, pp. 4-11.
44. *Daily Report: China*, Vol. 87-075, April 20, 1987, pp. K11-K14.
45. *Ibid.*, K12.
46. Gao Qinglin, "Implement the Plant Director Responsibility System in an All-Round Way—A Focal Point of China's Reforms in 1987," *China Report: Economic Affairs*, February 18, 1987, p. 23.
47. The following remarks are based on *ibid*, pp. 22-27.
48. *Ibid.*, p. 20. But even in the spring of 1987 state officials were pushing for more independent enterprise management. See Yuan Baohua, "Explore and Create Chinese-Style Socialist Enterprise Management," *Daily Report: China*, Vol. 87-072, April 15, 1987, pp. K7-K10.
49. "Chongqing Tries Management Responsibility System," *China Report: Economic Affairs*, February 27, 1987, p. 12.
50. "Chongqing Practices Asset Management System," *China Report: Economic Affairs*, January 6, 1987, pp. 17-18.
51. Li Changchun, "A New Attempt at Invigorating Large and Medium-Sized Enterprises," in US Foreign Broadcast Information Service, *China Report: Red Flag*, June 10, 1986, pp. 27-34. It is interesting that the CCP's major organ, *Red Flag*, carried the article urging that party cadres get behind and support the new industrial enterprise responsibility system.
52. Gao Shangquan, "New Trends and Tasks in Changing Economic Structural Pattern," *China Report: Red Flag*, April 14,

1986, pp. 22-31.
53. Wen Qinggui, "The Burden is Heavy and the Road is Long—Discussing the Position and Role of the Party Committee Secretary after Implementing the Factory Director Responsibility System," *China Report: Red Flag*, March 6, 1986, pp. 55-62.
54. "New Independent Enterprises Developed in Anhui," *China Report: Economic Affairs*, January 6, 1987, p. 16.
55. *Ibid.*
56. Dorothy J. Sollinger, "City, Province and Region: The Case of Wuhan," a paper presented at the conference, "China in a New Era: Continuity and Change," Manila, August 24-29, 1987.
57. *Ibid.*, p. 38.
58. *Ibid.*, p. 40.
59. Roy Grow, "Changing the Rules: Debating Price and Contract Regulations in the Northeast," a paper presented at the conference "China in a New Era: Continuity and Change," Manila, August 24-29, 1987.
60. For example of such corruption reported in the press, see *Renmin ribao*, February 4, 1986, p. 5; February 6, 1986, p. 1; February 7, 1986, p. 1; March 28, 1986, p. 1; April 15, 1986, p. 1; May 23, 1986, p. 1; and October 20, 1986, p. 1. "During the past several years, with the launching of Deng's reforms, criminality has been constantly on the increase. In the evening, people were afraid to go out and many crimes were going unpunished. According to official statistics, from 1980 to 1985 there were 40,000 murders in the whole of China, but the police were only able to catch 4,000 suspects," Tiziano Terzani, *Behind the Forbidden Door: Travels in Unknown China* (New York: Henry Holt and Company, 1984), p. 235.

SIX

RURAL FACTOR MARKETS IN CHINA
AFTER THE HOUSEHOLD RESPONSIBILITY SYSTEM REFORM

Justin Yifu Lin, Development Institute, Beijing

Between 1980 and 1984, China's agriculture attained the second most sustained overall expansion of the last three decades, second only to the period between 1963 and 1965—a recovery from the destruction of the Great Leap Forward.[1] The national income from agriculture grew at rates 7.5 percent (1981), 11.6 percent (1982), 9.6 percent (1983), and 14.5 percent (1984). This remarkable growth is mainly a result of the change from the collective system to the new household-based farming system.[2]

Before the recent institutional reform, a production team, usually consisting of about 30 neighboring households, was generally the basic unit of production and accounting in agriculture. The team was entitled to all factors of production. These factors were allocated under the unified management of a team leader with the exception of small private plots reserved for households' use in their spare time. Peasants, working under the supervision of a team leader, were credited with work points for a day's work that they had done. At the end of a year, net team income was first distributed among team members according to basic needs, then the rest was distributed according to the work points that each one had accumulated during the year. This institution was found to be very inadequate in providing work incentives to peasants in a production team.[3]

A new policy called the production responsibility system was introduced at the end of 1978 as one element of a package of reforms aiming at improving agricultural production in rural areas.[4] At first, this policy was designed to improve the management and incentive problems within a team. However, it developed into a specific form now called "the household responsibility system" that dissolved the production teams and restored individual households as units of agricultural production and accounting. The household responsibility system evolved into the main feature of the recent reforms in the Chinese rural areas. By the end of 1983, 97 percent of the production teams in China were converted to the new system. It is found that the shift from the production team system to the household responsibility system increased the agricultural productivity about 20 percent. This jump in productivity explained about 60 percent of the output growth in agricultural production between 1980 and 1983.[5]

As the impact brought about by the household responsibility system reform is a one-time discrete jump in productivity, the gain should have been exhausted.[6] However, the success of this institutional change prompted the Chinese government to push the market-oriented reform to its urban economy at the end of 1984. The urban economy is much more complicated. Any policy, good or bad, will not manifest itself in a short period. A policy essential for the long run may even cause great difficulties in the short run. Therefore, whether the market-oriented reform in urban areas will be persistently carried out may again depend on the performance of the rural reform.

In a normal situation, the sources of agricultural growth come mainly from the increases in inputs and technological improvement. From the historical and international experiences, we find that it is quite unlikely that the growth rate derived from the normal channels will achieve four percent per year.[7] However, even if four percent per year is achievable, it is not satisfactory. This rate is much lower than the annual average rate of 10.8 percent between 1980-1984. For the purpose of creating the consensus of supporting the overall market-oriented reforms in urban areas, an annual growth rate of six percent or higher in agriculture is necessary. To achieve such a high rate of growth, the hope lies in further reforms of rural economy.

The improvement in incentive resulting from the household responsibility system reform may have simultaneously created an allocative inefficiency. When the household responsibility system was introduced, land and other resources in a team were in most cases allotted to each household in proportion to its size. Therefore, for the households in a team, their land-person ratio was equalized after the household responsibility system reform. Households are at different stages in the life cycle. They thus have different endowments of family labor. In addition, households differ in abilities. An equal land-person ratio across households in a team thus does not fully equalize land-labor ratio across households. If each household faces the same production function, this egalitarian allocation of land will result in disparities in the marginal products of land and labor across households.[8] These differences in marginal products represent an allocative inefficiency. The higher than normal rate of six percent growth per year in agriculture may be achievable if resources are reallocated.[9]

One possible way to take advantage of these opportunities is through direct government intervention, like land-reallocation among households. Nevertheless, government intervention can be ruled out as an alternative for the near future. When the household responsibility system was first introduced, the land contracts in general ranged from one to three years. When an original contract expired, land was reassigned and adjusted according to changes in household size and labor endowment. This practice was soon found to be impractical. As land might be assigned away in the next contract, each household thus lacked incentives to invest in land improvement and to maintain properly the soil fertility. To overcome this disincentive in land investment and land maintenance, the Chinese government has adopted a policy of lengthening the contract of land usage to each household for up to 15 years or longer.

The other possibility for improving allocative efficiency is through market transactions. Market transactions can range from hired labor to land tenancy or may be packaged in complex contracts involving several transactions in different markets. Transactions in land and labor naturally will give rise to demand for credit. If factor transactions are costless, certain, unconstrained, and enforceable, then marginal products will be brought into equality by market transactions. However, as discussed by Binswanger and Rosenzweig, factor transactions in rural areas are characterized by risk and beset with incentive problems.[10] The existence of well developed rural factor markets cannot therefore, be taken for granted. This paper is devoted to examining the extent and possible developments in rural factor markets in China.

Before going into any detailed discussions, three specific features that characterize China's rural factor markets need to be mentioned:

a) The Chinese government has launched a second-stage

RURAL FACTOR MARKETS IN CHINA AFTER THE HOUSEHOLD RESPONSIBILITY SYSTEM REFORM

rural reform in 1984. The main theme of the second-stage reform is to transform a self-subsistence economy into a commodity production and exchange economy by way of readjusting the production structure in rural areas through market mechanisms. When the household responsibility system was first introduced, hiring labor, subleasing land, and lending money at high interest rates were all explicitly prohibited.[11] Since then there have been substantial changes. The first change came to the credit market. Private credit with a high interest rate is no longer categorically classified as usury in the 79th document issued by the State Council in 1981. Leasing out land to other farmers and hiring workers within a limited number (less than eight) were also formally sanctioned in Document No. 1, issued by the Central Committee of the Communist Party of China in 1984.

Transactions in factor markets have been legalized. However, socialist sentiment is still deeply rooted in China. It appears unlikely, for example, that the government will force a person to be evicted from his house if he uses it as collateral and fails to repay his loan. It is also unimaginable that public opinion will sympathize with the lender in the case of a default.

b) There is a commonly held belief in China that at least 30 percent of the labor force in rural areas is surplus labor. The argument is that the cultivated land per capita in 1949 was 2.7 *mu* and now it has shrunk to about 1.6 *mu*. However, the percentage of labor force in rural areas has been about the same during this period. If the surplus labor is defined as the labor force that can be removed from agricultural work in the peak period without reducing agricultural output, the accuracy of this belief is very doubtful. There has been tremendous investment in land improvement. The usages of chemical fertilizers and other modern inputs have also increased greatly. It is hard to imagine that the marginal productivity of labor in the Chinese rural areas could be zero or negative in the peak period. A more accurate way of expressing the situation would be that, under the current price system and the average operational landholding, the value of average product of labor in agriculture, especially in cropping, is much lower than that in non-agricultural sectors; therefore, there is a general tendency for the rural labor force to shift out of the agricultural sector. A study shows that the average net income per worker in the suburbs of Shanghai in 1981 was 441 *yuan* for agriculture, 1,003 *yuan* for sideline production, and 1,625 *yuan* for industry.[12] This income differential will thus induce a tendency for the labor force to move out of the agricultural sector.

c) The original production teams are still entitled to the ownership of land after implementing the household responsibility system. However, the use right of land is assigned to individual households for a period of 15 years or more. This practice created a situation very similar to the distinction between "topsoil right" and "subsoil right." The "subsoil right" represented the usual, original claim to land ownership, including the right of sale, but excluding the right of cultivation. The "topsoil right" was the right to cultivate a piece of land, which could also be leased or sold. Hence, in fact, two distinct rents, one for the subsoil right and the other for the topsoil right, were involved.[13] The entitlement to a use right of land for 15 years, therefore, is a sufficient condition for land market transactions.

LAND ALLOCATION AND LAND MARKET

The differences in the marginal products of land and labor in China have two major sources. One is the difference in land endowment across regions. The other one is the egalitarian distribution of land after the household responsibility reform.

Table 1A shows that the eight provinces that have the lowest land-labor ratio possessed 39.6 percent of the total labor force in China in 1983; however, they only had 21.4 percent of the total cultivated land. One the contrary, the nine land-rich provinces possessed only 10.5 percent of the total labor force but were endowed 34.1 percent of the total cultivated land. The peasants in land-rich Helongjiang Province on the average had about 17 times as much land as the peasants in land-poor Zhejiang Province. Not only is the distribution of cultivated land unequal among provinces, but it is also unequal within a province. Table 1B shows that, in Anhui Province in 1983, 31.8 percent of cultivated land located in nine prefectures had 22.9 percent of the labor force. In contrast, the six prefectures that had the lowest land-labor ratio had 21.7 percent of the labor force but only 13.6 percent of the cultivated land. Although the differences within Anhui Province are not as large as the differences among provinces, the disparities are still quite substantial. The peasants in Huaibeishi have 3.6 times as much land as the peasants in Anqingshi. The differences in land-labor ratio reduce after adjusting for irrigation (proxy for land quality) and multiple cropping (proxy for climate and temperature). However, the differences are still very substantial as the last columns of Tables 1A and 1B suggest. Although, without empirical studies, it is difficult to say to what degree land endowments differ across neighboring production teams, the difference itself can be taken as a fact. The distribution of inherited intelligence of a large population approaches normal in any large sample. There is no prior reason to believe that the average quality of the labor forces in two neighboring teams, which both have about a hundred workers, would be significantly different.[14] It thus should not be too unrealistic to assume that the quality of labor forces across teams and regions is the same. Consequently, much of the differences in the land-labor ratio represents an allocative inefficiency.

Allocative inefficiency within a team, however, would arise

Table 1A
LAND ENDOWMENT IN EACH PROVINCE

Province	(1) Labor	(2) Cultivated Land	(3) Land-Labor Ratio	(4) % of Area Irrigated	(5) Multiple Cropping Index	(6) Effective Land-Labor Ratio
Guizhou	10,087	28,480	2.8	24	153	3.8
Sichuan	38,712	98,109	2.5	47	181	3.9
Zhejiang	14,030	27,249	1.9	84	252	4.0
Yunnan	12,808	42,488	3.3	34	140	4.3
Guangdong	19,061	47,130	2.5	65	200	4.4
Guangxi	13,963	39,301	2.8	54	177	4.4
Fujian	7,083	19,240	2.7	64	189	4.5
Tibet	825	3,437	4.2	53	93	4.6
Hunan	20,239	50,998	2.5	82	218	4.8
Henan	25,370	106,508	4.2	45	160	4.9
Shanghai	2,071	5,249	2.5	98	218	4.9
Anhui	17,478	66,518	3.8	50	177	5.9
Jiangsu	20,068	69,451	3.5	75	184	5.9
Shandong	24,988	107,728	4.3	63	146	6.1
Hubei	14,572	55,481	3.8	63	200	6.6
Tianjin	1,391	6,879	4.9	67	133	6.7
Beijing	1,372	6,343	4.6	81	151	6.9
Hebei	17,481	98,551	5.6	54	131	7.3
Jiangxi	9,305	35,753	3.8	74	229	7.4
Shaanxi	9,124	56,377	6.2	33	127	7.6
Qinhai	1,104	8,640	7.8	27	87	7.7
Liaoning	6,488	54,814	8.4	19	102	8.9
Gansu	5,841	53,425	9.1	24	98	9.5
Shanxi	6,647	58,075	8.7	28	107	9.6
Ningxia	1,061	12,605	11.9	28	101	12.8
I. Mongolia	4,657	75,974	16.3	20	91	16.3
Jinan	3,845	60,895	15.6	18	100	16.3
Xingjiang	2,570	47,422	18.5	83	92	21.4
Heilongjiang	4,110	131,273	31.9	7	98	32.1

Source: *China Agriculture Yearbook 1984.*

Note: (1) Agricultural labor force excluding workers in village-run industry, unit = 1,000 workers; (2) cultivated land unit=1,000 mu; (3) col. 2/col.1; (4) % of cultivated land irrigated; (5) unit=%; (6) effective land-labor ratio is the land-labor ratio adjusted for irrigation and multiple cropping; its formula is: effective land-labor ratio=Land-labor ratio x (1 + % of area irrigated/4) x (1 +(Multiple cropping index- 100)/2). See A. M. Tang, *An Analytical and Empirical Investigation of Agriculture in Mainland China 1952-1980* (Seattle: University of Washington Press, 1984) for the rationale of these adjustments.

Table 1B
LAND ENDOWMENT IN EACH PREFECTURE, ANHUI PROVINCE

Prefecture	(1) Labor	(2) Cultivated Land	(3) Land-Labor Ratio	(4) % of Area Irrigated	(5) Multiple Cropping Index	(6) Effective Land-Ratio
Anqing	85	125	1.5	88	215	2.9
Huizho	705	1,192	1.7	76	248	3.5
Anqing	2,094	4,859	2.3	66	211	4.2
Tongling	124	297	2.4	95	200	4.5
Wuhu	697	1,736	2.5	77	224	4.7
Maanshan	297	756	2.5	93	225	5.0
Chaohu	1,527	4,355	2.9	80	199	5.2
Liuan	2,004	6,706	3.3	68	171	5.2
Fuyang	4,363	16,915	3.9	24	166	5.5
Xuancheng	764	2,246	2.9	65	222	5.6
Hefei	1,172	4,183	3.6	72	172	5.7
Huainan	390	1,522	3.9	57	174	6.1
Suxian	1,669	7,913	4.7	29	158	6.5
Huaibei	416	2,208	5.3	11	160	7.1
Chuxian	1,297	6,250	4.8	68	165	7.4
Bangbu	844	4,542	5.4	47	162	7.9

Source: Statistical Bureau, Anhui Province.

Note: Definitions and units are the same as in Table 1A, except labor force here includes workers in village-run industry. Data are for 1983.

from an opposite reason. Under the production team system, the team-owned land was divided into collectively farmed plots and private plots. Private plots were allotted to each household according to its size. The land that could be allotted for private plots varied from time to time. The average amount of land in private plots nationally was 5.7 percent in 1978. It rose to 7.1 percent in 1980.[15] After the introduction of the household responsibility system, the collectively farmed land was contracted to individual households in two different categories. One was the "food ration plot." The other one was the "responsibility plot." The difference between these two kinds of plots was that a household had to pay only state tax on the food ration plot, but it also had to pay the public accumulation fund, public welfare fund, and other duties to its team on the responsibility plots. As for the private plot, the state tax was also waived. Two different practices were used to contract the collectively owned land. The first practice contracted the land strictly in proportion to the size of each household. The second one took into account both the size and the labor force of each household. However, the results of these two practices may not be very different. A survey of a production team in Guangxi Province found that the household with the largest labor force only had 0.16 *mu* per capita more than the average of the team, and the household with the smallest labor force had only 0.078 *mu* per capita less than the average of the team, even though 70 percent of weight was given to the labor force in the contracts.[16] Therefore, it can be assumed that the land-person ratio across households in a team is roughly equal no matter what practice has actually been adopted. Not only is the quantity of land per capita equal across households, but the quality of land owned by each person in a team is also the same. This is because land was first graded according to its quality, then each person received a piece of land from each grade. Therefore, each household in China after the individual household reform often owns more than 10 strips of land.[17] Households in a team are at different stages of their life cycles and thus have different labor endowments. They also have different levels of education, experience, and other abilities. As a consequence, the equal land-person ratio across households in a team generates a potential allocative inefficiency. A survey of 235 households in a village in Sichuan Province found that 25 percent of the households with a rich labor endowment did not have enough land to farm; 6 percent of the households did not have enough labor to work on their land; and 4.7 percent of the households were good at other trades, so they did not want to work on their land.[18]

Land transactions in China's rural areas after the

household responsibility reform are restricted in their form of lease. The government has encouraged the households specialized in cropping to consolidate their landholding.[19] Table 2 summarizes several studies concerning the extent of land transactions in certain regions in China. Rows 1-3 are based on the surveys done at the end of 1983; rows 4-5 are based on the data collected at the end of 1984. Column 2 is the percentage of households in an area that either leased out their land to other households or returned their land to their production teams. The land returned to a production team may be recontracted to other households. Column 3

Table 2
SCOPE OF LAND TRANSACTIONS IN SOME AREAS

	Household involved (%)	Land involved (%)
Xiapu County Fujian Province[a]	4.0	2.6
Ezhoushi Hubei Province[b]	5.0	3.0
Huangni Township Chuzhou, Anhui Province[c]	12.2	3.3
Zhongwei County Ningxia Province[d]	0.058	0.025
Tianjin[e]		8.3

Source: (a) Genxing Fan. "A Survey of the Approaches Used in Recontracting Land," *Fujianq luntan* 7, 1984, pp. 45-46;

(b) Xinglong Wang. "On Current Stage of Land Recontracting in Rural Areas," *Honqqi* 8, 1984, pp. 24-28;

(c) Changmin Hou and Tanghou Dou. "Permitting Land Transfers Is Necessary for the Development of Productivity in Rural Areas," *Jianghuai luntan* 4, 1984, pp. 5-10;

(d) Nong Wei. "A Survey of the Concentration of Land to the Farming Expert in Zhongwei County," *Ninqxia shehuikexue* 1, 1985, pp. 84-86;

(e) Agricultural Department, Communist Party of China, Tianjin. "A Survey of the Situation in the Concentration of Land," *China Agriculture Yearbook, 1985*, pp. 409-11.

shows the percentage of land in an area that was involved in land transactions. The percentage of land involved is less than the percentage of households involved. This is due to the fact that most households only leased out or returned their responsibility plots and kept their food ration plots and private plots.

All these studies mentioned found that land transactions were more active in areas closer to cities. Tianjin is the third largest city in China. But even by the end of 1984, altogether only 8.3 percent of land had been leased out in Tianjin; therefore, land transactions in China as a whole must have existed with only a very limited scope.

These studies also found that the majority of households that leased out or returned their land were "specialized households" that engaged in noncrop jobs, such as transportation, repairing, food processing, other services, or fish, poultry, and pig-raising. Only a very small portion of households leased out or returned their land because of lack of labor endowment. From the supply side, we find that the scope of land transactions crucially depends on the job opportunities outside cropping.

Although the extent of land transactions is very limited, the forms it takes are more extensive. They can be classified into two basic forms: (a) without compensation or (b) with compensation.

In the first case, households either give their land back to production teams or give it to their relatives or close friends. In either situation, households still maintain their claim over the use right to the land. They can take it back in the future if they desire. Rent over use right is positive (see the discussion later). Households voluntarily give up the rent entitled to them. This fact implies that (1) the land market in these places must not have existed, so the households that want to migrate out of agriculture could not find other households to lease it and (2) that the labor market or the credit market had also failed, so the households could not find workers to farm their land or did not have enough cash to hire workers.

For the cases with compensation, there are two main varieties: (1) rental and (2) sharecrop. Fan reported that in Fujian Province there were three ways in which rent was paid.[20] In one case the households that leased out land were guaranteed the right to purchase a certain amount of food grain at the government procurement price. Because the government procurement price was lower than the market price at the local fair, the difference between these two prices became rent. Fan found that rent paid in this way was equivalent to 64.86 *yuan* per *mu*. In the other case, the households were compensated with a given amount of free grain, ranging between 200-300 *jin* of grain. Fan found that the market value of it was about 60 *yuan* per *mu*. In still another case, rent was paid in cash at also about 60 *yuan* per *mu*. In all these cases the rent was about 30 percent of the gross value of output. Fan also reported a case of sharecropping. A bee-raising specialized household leased out its land of 5.2 *mu* and lent 300 *yuan* to the renter for the cost of seeds and fertilizer. The renter harvested 5,600 *jin* of rice. For the required quota, 1900 *jin* were sold for 320 *yuan* to the government. This money was paid back to the landholder for the 300 *yuan* loan. The rest of the 5,600 *jin* were equally shared by the landholder and the renter. The rent amounted to 129.27 *yuan* per *mu* according to the market value of rice at the local fair. In the other study of a county in Zhejiang Province, Zhou and Du found that fixed rent was paid in two ways.[21] The rent was equivalent to 57.7 *yuan* per *mu* when a household was guaranteed the right to purchase a certain amount of grain at the government procurement price. It was about 52.5 *yuan* per *mu* at the local fair price for rent in the fixed amount of free grain. The rent was also roughly about 30 percent of the gross value of output. Zhou and Du found that there was a tendency to use the rent in the fixed amount of free grain. They also recorded a case where a household hired casual workers to farm its land. The net income per *mu* for the landholder in this case was 77.97 *yuan*. That was about 30 percent higher than the prevailing rent.

There are several interesting relations in these cases:

a) Rent in cash was a little bit lower than rent in kind. This may be explained by the facts that cash is preferred because of its general purchasing power and that the price for grain at local fairs may fluctuate, so there is some risk inherent in rent in kind.

b) Among the rent in kind, the rent was lower if it was paid by a fixed amount of free grain than if it was paid by a fixed amount of grain at the government procurement price. This again may be due to the fact that the landholder has to face larger risk because of the possibility of price fluctuation at the local fair.

c) The return to land was higher for a landholder if he hired workers to farm it instead of leasing it. This can be explained by the fact that a landholder has to face the risks arising from production and market fluctuations and that he also contributes his entrepreneurship to production.

d) The land market is tied with the credit market in the case of sharecropping, as reported above. The return from leasing to the landholder depends on how the interest rate is calculated. In Chinese rural areas the interest rate is extremely high for private credit. It ranges between 4-10 percent per month or even higher. Because the interest of the loan to the renter was not explicitly paid, after deducting the implicit market interest rate, the rent for the sharecropping case was not as high as it appeared to be.

In the Chinese rural land market, a long-term lease with advanced payment of the present value of all rent for the use right of land for 15 years has not been found. For a household leaving agriculture to establish a noncrop business, this kind of transaction should be attractive. It is a good way to overcome the possible cash constraint for starting a business. The lack of such transaction can only be understood from the demand side. There are two possible explanations. One is that land cannot be used as a collateral because the government will not enforce a lender's right by helping him repossess the land in case of default. In general, the cash-rich households in China are households specialized in non-crop activities. From the above discussions, we find that they are the households that would like to leave their land. The households that like to expand their landholdings are, in general, those households that stay in the crop sector. They are more likely to be cash-poor. Therefore, unless they can borrow from a credit market, they will not be able to finance land buying. Credit markets will be limited if land cannot be used as collateral.[22] Therefore, a cash-poor household will not be able to finance the transaction if it cannot use the purchased land as collateral. The other explanation is that the government will not protect such a contract, if the leasing household fails in nonfarm undertaking and tries to take its land back. Therefore, even if a household is not constrained by a cash requirement, it may be still reluctant to expand its landholding through such a transaction.

RURAL LABOR MARKETS

Transactions in labor are another way to equalize differences in marginal products across regions and households. Labor-hiring was prohibited before the recent reform. When labor transactions are prohibited, migration between teams and across regions can be another way to bring marginal products into equality. However, as rent was suppressed in a collective system, workers were compensated with the average net product instead of the marginal product. Consequently, a portion of their income actually was rent. Workers in a team with a lower average income certainly have the incentives to migrate to a team with a higher average income. They would be able to receive the same higher average income as the original members in the higher-income team. Nevertheless, the workers in a higher-income team would be reluctant to accept migrants from other teams for fear that their rent would be shared by the newcomers.[23] Therefore, when payment of marginal product to workers was prohibited, the migration between production teams or across regions was virtually non-existent.

When the household responsibility system was first introduced, hired labor was explicitly prohibited on the grounds that exploitation of the surplus value was not allowed to be restored.[24] Nevertheless, labor-hiring can be mutually profitable for both the employers and employees. With such underlying incentives, it is difficult to enforce the decree. As more and more cases of hired labor appeared and the government realized labor transactions were beneficial for the economy as a whole, the policy was revised to allow hiring labor. Yet a household is limited to hiring not more than eight workers. The limit of eight workers is chosen because once Marx wrote in *Das Kapital* that a person who hired less than eight workers could not be classified as a capitalist, as he still had to attend to the physical work himself. The arbitrarily set limit has never been strictly abided by. Some households in rural areas have hired more than a hundred permanent workers.[25] While the upper limit of eight workers is still officially maintained, the government does not seem to enforce it.[26]

The opening of labor markets makes the equalization of marginal products across households possible by way of labor transactions. What is of interest is to what extent the difference in marginal products has been narrowed. A survey of labor-hiring in Wu County, Jiangsu Province, found that more than 50 percent of the labor hired was used in nonagricultural work (see Table 3). In Wu County the majority of labor was hired for civil engineering or manufacturing. The major impact of opening labor markets in Wu County is thus the increase in job opportunities within the non-farm sector. Another survey of labor hired in Yangshi County, Shenyangshi, Liaoning Province has the same finding (see Table 3). The impact of labor-hiring on narrowing the differences in marginal productivity across

Table 3
LABOR-HIRING IN WU COUNTY AND YANGSHI COUNTY

	No. of household	No. of workers hired(%)
Wu County[a]		
Total	356	2073 (100)
Civil engineering	145	842 (40.6)
Manufacturing	112	651 (31.4)
Retailing	52	303 (14.6)
Stock-raising or farming	8	52 (2.5)
Transportation, fish-raising, and others	39	225 (10.9)
Yanqshi County[b]		
Total	105	970 (100)
Civil Engineering	5	277 (29.9)
Manufacturing	27	143 (14.7)
Retailing	6	19 (2)
Farming	48	439 (46)
Transportation	19	92 (18)

Source: (a) Songmao Zhang, "A Study of Labor-hiring in Rural Areas," *Jianghai xuekan*, 5, 1985, pp. 42-44; (b) Xuechen Gao and Guozhi Lu, "Adhering to the Orientation of Socialist, Cooperative Economy and Actively Guiding a Healthy Development of All Forms of Economies," *Nonqye jingji (Shenyang)* 5, 1985, pp. 15-20.

households is ambiguous for the workers hired for non-farm jobs. Non-farm jobs often require special talents; hence, from the supply side, the labor working for nonagricultural jobs is not necessarily coming from households with more labor endowment. However, from the demand for agricultural workers, three kinds of households may hire workers: (a) households specializing in stock-raising, fish-raising, or vegetable cultivation; (b) households renting large amounts of land; and (c) households keeping some of their labor force at home for farming but having shifted the major part of their labor force out of agriculture. The first two categories indicate that the labor-hiring households have superior technology or entrepreneurship in agricultural productions. The last one indicates that the remaining labor endowments in the labor-hiring households must be less than the average. From the demand side, therefore, transactions in labor market tend to reduce the differences in marginal products across households.

The first two categories of households usually hire workers on a monthly or yearly basis. Zhang Songmao reported that a household in Wu County rented 430 *mu* of land from its own and neighboring counties and employed 18 workers for producing grain, watermelon, soybeans, and so on. It also raised 550 chickens and ducks. The head of this household was formerly a production team leader.[27] Zhou found that a household in Hainan, Guangdong Province, rented a 300-*mu* sugar cane plantation and hired 20 workers to run it.[28] The last category of households usually hire casual workers either by piece rate or day rate. The wage rate for a permanently hired worker was about 1000 *yuan* per year in both Zhang's and Zhou's studies. For the casual workers, Zhou and Du found that the wage rate was about five *yuan* per day in the peak period and about four *yuan* per day in the off-peak period in 1983 in Zhejiang Province.[29] Shi found that the piece rate in the suburbs of Shanghai in 1979 was 15 *yuan* to 17 *yuan* for transplanting a *mu* of rice seedlings. It was equivalent to 2.5 *yuan* to 2.8 *yuan* per day.[30]

As discussed before, it was uncommon for a production team to be willing to accept a migrant from the other teams or from the other regions before the transactions in labor were legalized. Therefore, another natural impact of opening labor markets was the migration of labor across regions. A study found that by 1984 over a thousand workers from other provinces had been employed permanently in the suburbs of Shanghai. Some of them worked in the village-run industries. However, a substantial portion of them worked in vegetable gardening, duck-raising and chicken-raising. A brigade was found to have hired 85 migrant workers in 1984; it had planned to hire 50 more in 1985. The migrant workers would constitute 56 percent of the labor force in this brigade by 1985.[31]

Labor markets in China's rural areas are still very limited.[32] Furthermore, only a portion of labor hired in rural areas is actually engaging in agricultural work. From the characteristics of households that hire workers for agricultural work, we find that none of them are households with the least family labor endowment before any market transactions. From my observations, most households with the least labor endowments solve their problem by growing crops with different harvest periods, so the demand for labor at each peak period is mitigated. If there are still shortages of labor in some period, they engage in direct labor exchange with either neighboring households or relatives and friends. The reason for this may be that hired labor is subject to incentive problems. The direct exchanges of labor between relatives and friends mitigate the shrinking problem and thus reduce the cost of supervision. The other explanation is that hiring labor requires cash. Only households with good access to credit markets or high cash income have the ability to hire workers. As credit markets in rural areas are not developed, households with the least labor endowments will not have good access to credit markets. They are obviously not households with high cash incomes. Therefore, households with the least labor endowments may be unable to finance labor-hiring because of the cash constraint.

RURAL CREDIT MARKET

In the above two sections, it is found that the limited extent of rural land and labor markets may be closely related to the limited rural credit markets. The subsequent discussions focus on the extent and constraints of credit markets in Chinese rural areas.

In a socialist society, there is a strong sentiment against the taking of interest. In Marx's teaching, interest in a capitalist society is a redistribution of the exploited surplus value between financial capitalists and industrial capitalists. However, the seasonality of agricultural production gives rise to seasonal needs for funds to bridge gaps between receipts and expenditures. Both formal and informal credit markets existed in rural areas even before the household responsibility system reform (see the discussions following).

There are no private financial institutions in China. Formal credits are provided by the Chinese Agricultural Bank and credit cooperatives. The Chinese Agricultural Bank is a state bank. It has branches in every commune. Credit cooperatives are formally owned by commune members. However, credit cooperatives in the past, in reality, acted as branches of the Agricultural Bank. In many areas, the Agricultural Bank and credit cooperatives shared the same offices and had the same staff.[33] The credits were provided at subsidized interest rates in the past. The interest rate charged for a loan was 0.25 percent per month until 1981. However, the average interest rate for deposit was 0.31 percent in 1980. The government in Shanxi Province thus had to subsidize 11.5 million *yuan* for the Agricultural Bank and the credit cooperatives in 1980 alone.[34] Not only were the interest rates charged low, but loans were often provided without consideration of their prospects of recovery. For example, in Shanxi Province, only 84.6 percent of loans between 1976 and 1979 were paid back.[35] The situation was not better in the other provinces. According to national statistics, four billion *yuan* of bad agricultural loans were cancelled in 1961 and another eight billion *yuan* of bad loans were accumulated between 1962 and 1980.[36] The availability of credit was thus severely limited. A survey of several counties in Henan Province found that, due to poor recovery, each county had only about two million *yuan* for new loans although each of them were officially allotted more than ten million *yuan* for agricultural loans.[37]

The results of low interest rates and low pressure for repayment are not hard to figure out. Credits were not used with care. For example, in Linfen county, Shanxi Province, 140,000 *yuan* of agricultural loans from credit cooperatives before 1978 were not used properly. Among these loans of 140,000 *yuan*, 100,000 *yuan* was used on construction that was never completed; 20,000 *yuan* was expended on unusable materials; and 20,000 *yuan* was wasted on administrative expenditures.[38] The other result is credit rationing. As interest rates were low and pressures for repayment were not strong, real opportunity costs for using credits were close to zero or even negative. Therefore, the demand for credits was definitely higher than the supply of credits. The market could thus not possibly be cleared without non-market measures. The criterion for rationing varied from time to time. Sometimes the priority was to help poor teams. At other times the priority was given to rich teams that had better uses for the funds.[39] However, it was often found that a county leader used ad hoc criteria in deciding who should be given a loan.[40]

The availability of formal credit declined because of the accumulation of bad debts and because of the unwillingness of people to deposit in credit cooperatives.[41] However, the demand for credits increased sharply. Taking Gansu Province as an example, total agricultural income increased 107 percent between 1956 and 1979, yet production expenditures increased 278 percent. In 1956, expenditures consisted of 21 percent of gross income; it increased to 39.2 percent in 1979. The situation in other provinces was no better. Statistics involving 3.6 million production teams in 26 provinces, provided by the Agricultural Bank, showed that in 1980, on the average a production team had only 15 percent of the required working funds. Another survey showed that about 40 percent of the production teams in China did not have any working funds at all. Since formal credits could not satisfy the need for working funds, many production teams had to rely on private credits. The same report showed that in some regions as much as 70 percent of the production teams engaged in informal credit markets.[42] The interest rates paid for private credit were extremely high in some areas. One such case was recorded in a study of Dancheng County, Henan Province. The study found that the combined revenues for production teams was 21.76 million *yuan* between January and September 1979, yet the expenditure was 36.33 million *yuan*. Half of the deficit was financed by loans from the Agricultural Bank and credit cooperatives. The other half was borrowed from private sources. Seventy percent of the credit borrowed from private sources was used to purchase fertilizer, 20 percent was used to buy livestock, the remaining 10 percent was paid for administrative expenditures. The interest rates ranged from 3 percent per month to 30 percent per month. On the average, it was 10 percent per month.[43] Zhang in another study, however, showed that the interest rates only ranged between 2-5 percent per month.[44] Private credits were mainly provided by members in the borrowing production team. Among the 256 people lending money to a production team in Miluo County, Hunan Province, Zhang found that 235 were members of this team, 18 were cadres and government staff (not team members), and 3 were urban residents.

Before the household responsibility reform, the majority of credits from the Agricultural Bank and credit cooperatives were given to production teams. Borrowing and lending between individual households were also rare. A study found that of the 3.03 million *yuan* private credit in Xingmin County, Liaoning Province, January through May 1980, 1.3 percent was between state firms and production teams, 3.3 percent was among different production teams, 93.7 percent was between production teams and individual households, and only 1.7 percent was among different individual households.[45]

The individual household responsibility reform brought dramatic changes in rural credit markets. In the production team system, an individual household would not need credit for production purposes. If a household had emergency needs for consumption, health, marriage, and so forth, its production team was more or less obliged to take care of them. The loans from a production team to individual households, in general, were interest free and were not required to be paid back until the households were able to do so. The household responsibility system restored the individual household as the basic unit of production and accounting. It also eliminated the group insurance provided by the production team system. Therefore, individual households became the primary actors both in formal and informal credit markets. For the nation as a whole, among the 16.6 billion *yuan* loans from credit cooperatives in 1983, 46 percent were given to individual households. The figure was only 19.6 percent in 1980.[46] The actual new credits to individual households should be higher than this figure suggests because many of the loans to the collectives were old loans that had not been repaid.

The other new feature in rural credit markets after the household responsibility reform is the sharp rise of the amount of cash in circulation. This is partly because of a marked price rise for government-purchased agricultural products in 1979 and partly because of remarkable output growth since 1978. This feature is reflected in the dramatic increase in deposits in credit cooperatives. The deposits in credit cooperatives by individual households were 7.8 billion *yuan* at the end of 1979. This figure rose to 32.0 billion *yuan* at the end of 1983. Most of the increased deposits were redeposited in the Agricultural Bank as reserve. Between 1978 and 1983, deposits in the credit cooperatives increased by 6.46 billion *yuan* annually; however, loans from credit cooperatives increased only by 2.42 billion *yuan*. It is suggested that the fact that loans increased by less than deposits was due to the inertia of credit cooperatives.[47] However, it may be due to the fact that a local rural financial institution has to keep a high reserve ratio in order to prevent illiquidity. The seasonality of the agricultural production leads to synchronic timing of deposits and withdrawals. Covariance of yield risk leads to covariance of default risk. Therefore, a local financial institution has to keep a high reserve ratio to keep liquid.[48]

As production expanded after the individual household responsibility reform, the demand for working funds also increased sharply. A survey of 21 households located at Xiachai Village, Ningdu County, Jiangxi Province, carried out by Mei[49] found that per capita cash income was 59.25 *yuan* in 1978 and 209.66 *yuan* in 1983. Meanwhile, the total money borrowed in the sample was 415 *yuan* in 1978 and 3,870 *yuan* in 1983. The weight of borrowed cash in total cash income was 7.4 percent in 1978 compared with 20.5 percent in 1983.

As the availability of formal credit was limited, private credit was the major source of rural credits. Mei reported that in Ganzhou Prefecture and Jiujiang Prefecture, both of Jiangxi Province, private credit was twice as much as the credit from the Agricultural Bank and credit cooperatives. Zhang Zhiping[50] making another survey of 20 households in two counties in Helongjiang Province, had the same findings.

Table 4
A SURVEY OF CASH EXPENDITURE AND TYPES OF CREDIT

County	Households Surveyed	Cash Expenditure on Inputs (*yuan*)	Bank and Credit Cooperative	Credit Private
Niujia	10	8,367	2,056	2,880
Chonghe	10	5,241	35	2,050

Source: Zhang Zhiping, "A Survey of the Rural Private Credits in Wuchang County, Heilongjiang Province," *Jinqji wenti tansuo* 7, 1985, pp. 40-42.

Zhang's survey is summarized in Table 4.

Zhang Zhiping also found that about 40 percent of private credit was used to buy draft animals, tractors, chemical fertilizer, and pesticides for grain production; about 50 percent was used for expanding cash crops, husbandry, and other production; and about 10 percent was used for repaying matured loans and for consumption. The terms of credit for grain production in general did not exceed 10 months and for other production did not exceed a year. Private credit was obtained (a) from relatives or close friends; (b) from neighbors in the same village, directly or through middlemen; and (c) from residents in other counties through middlemen. All private credits in rural areas depended on oral agreements. No explicit contracts were written. Furthermore, no collateral was found in private credit. The interest rates charged depended on the relationship between borrowers and lenders, credit-worthiness of borrowers and middlemen, and the expected returns of investment. The rate was about 3 percent per month for loans between close friends and about 5 percent per month for others.[51] Mei, however, found that in some cases the interest rates were as high as 10 percent or even 15 percent per month.[52]

While a 3-5 percent interest rate per month for private, short-term agricultural production loans is not uncommon in other developing countries and existed before the revolution in China's rural areas,[53] the interest rates were much higher for other types of private credit. Liu and Liang reported that the interest rates faced by private enterprises ranged between 4-10 percent per month; some were even as high as 20 percent per month.[54] In a society where high interest rates have been condemned for so long, it is an interesting phenomenon that private interest rates could be so high.

A feature that merits special attention is that, in all the studies mentioned above, no collateral is found in private lending and borrowing. While collateral is also not used very often in informal rural credit markets in other developing countries, the complete nonexistence of collateral is unusual. If collateral is used, the interest rate charged will be smaller because the risk of default declines as the value of collateral increases. The lack of collateral will also severely limit the extent of the credit market from both the supply side and the demand side. Lenders have to charge high interest because of the risk of default. However, the risk of default is a function of, among other things, the interest rate charged and the loan size. The expected gain for lenders may go down as the interest rate increases. Therefore, even though a borrower is willing to pay high interest rates, he may not be able to find someone to borrow from. From the demand side, the market may also disappear because the higher the interest rate is, the harder it is to find investment opportunities that have high enough expected returns.[55] Where loan sizes are large it would be beneficial for both borrowers and lenders to utilize collateral. In China's rural areas, there is no lack of private property that can theoretically be used as collateral. Houses are always privately owned. After implementing the individual household responsibility system, tractors, pumps, mills, trucks, draft animals, and livestock are also all owned by individual households. Furthermore, the use right of land that lasts 15 years or more is a property that could be traded and therefore serve as collateral. The lack of collateral in China's rural credit markets can only be explained by the reluctance of lenders to accept it.[56] Although the interest charged for private credit is legalized in China, the ideology is still strongly unfavorable toward lenders. Public opinion will definitely sympathize with a borrower in the case of unintended default. The Chinese government is not prepared to enforce lenders' rights by evicting the borrowers or assisting the lenders in repossessing the assets in the case of defaults. The absence of collateral in China's rural credit markets is consistent with Binswanger and Rosenzweig's thesis that the collateral value of an asset depends on the legal environment.

Because of the lack of use of collateral, several forms of tied contracts appear in rural China, as in countries where suitable collateral does not exist. A landholder may provide credit to his tenant at a very low interest rate as part of a land contract, as mentioned in the discussion of land markets. A third party guarantee is also often seen when borrowers and lenders are not relatives or close friends. It is also found that some private enterprises require new employees to invest in the firms as a precondition for hiring.[57] However, the most powerful guarantee for a lender in China's rural areas may be the threat of losing future borrowing opportunities when a borrower does not repay the loan. The rural population in China is relatively immobile. Information on default will be transmitted quickly to all potential lenders. Because insurance is absent in rural areas, access to credit provides an important substitute for insurance. Therefore, the loss of future borrowing opportunities is a very high cost for any borrower.

CONCLUDING REMARKS

Most barriers for factor market transactions which existed before the household responsibility system reform have been cleared. Land can be leased out for rent. Interest can be charged for credit. Labor can be hired with a limitation that is not enforced. However, transactions in land, labor, and credit markets are still not very active. The main reason may be due to the fact that a lender's right is not protected by the government; therefore, lenders have to charge high interest rates to offset the risks of default.

On the one hand, the average landholding in China's rural areas, in general, will not produce an income comparable to the income from other sectors. Therefore, there is a general tendency for a household to shift out of the agricultural sector, especially cropping. On the other hand, the egalitarian allotment of land after the household responsibility reform provides a safe shelter for every household. Unless a household is secured with a job that produces an income higher than cropping, it will not render its land to the other households. Because there also exists a labor surplus in urban areas, it is almost impossible for rural labor to find jobs in state or collective enterprises there. Outmigration from the cropping sector will be possible only if a rural household starts its own non-cropping business (such as fish-raising), non-farm business (such as transportation), or finds a job in rural private enterprises that have emerged after recent reforms. Limited credit at very high interest rates will greatly reduce the possibility of profitable private businesses for households with a relatively poor cash endowment. Therefore, a limited credit market may result in a limited outmigration from cropping and, therefore, land markets are limited from the supply side. Labor markets may depend on credit markets almost in the same way as land markets. The average landholding is very small. Most households do not have enough land to farm. Therefore, unless a household rents additional land from households moving out of cropping or has its own major labor force moving out of cropping, it will not hire workers for agricultural work. Hence, no matter if labor is hired for agriculture or non-agriculture, labor markets will be thin if non-farm and non-crop job opportunities are limited by credit markets.

If the government changes its position on lenders' rights in the case of default, it is predictable that the supply of credit will increase and interest rates will reduce. The opportunity for profitable business outside cropping will thus expand. The threshold for outmigration from the crop sector becomes easier to overcome. As a consequence, the scopes of outmigration from cropping, land markets, and labor markets, and the possibility of resource allocation through factor markets will all increase. Nevertheless, the government may have to tolerate the emergence of a landless population. If lenders' right is protected by the government, then among other things, the use right of a piece of land for 15 years will become acceptable as collateral. Since foreclosure implies loss of access to land, protection of lenders' right may lead to some households becoming landless. At present the government does not seem to be willing to undertake a policy with such consequences. However, rent, hired labor, and interest were all not acceptable to policy-makers a few years ago; they are all legal now. Therefore, it is not unimaginable that in a few years the Chinese government may change its position on enforcing lenders' rights.

CHINESE ECONOMIC POLICY

NOTES

The research for this paper is funded by the Agriculture and Rural Development Department of the World Bank. The author has benefited greatly from comments by Hans P. Binswanger, Robert F. Dernberger, Gershon Feder, D. Gale Johnson, Nicholas Lardy, Alan Piazza, Louis Putterman, Bruce L. Reynolds, Umnuay Sae-Hau, Theodore W. Schultz, Terry Sicular, Thomas B. Wiens, and the workshop participants at Yale University, PWPA, and Chinese University of Hong Kong. None of them bears any responsibility for the views presented and errors remaining in this paper.

1. The national income from agricultural output grew at an average rate of 11.5 percent per year between 1963 and 1965 (National Income Balance Statistics Division), National Bureau of Statistics, *Summary of National Income Statistics*, 1949-1985, p. 12. The growth rates between 1978 and 1985, reported here, are taken from the same source.
2. See Justin Yifu Lin, "Household Farm, Cooperative Farm and Efficiency: Evidence from Rural De-Collectivization in China," Discussion Paper No. 533 (New Haven: Economic Growth Center, Yale University, 1987).
3. For a detailed discussion of the incentive problem in a production team, see Justin Yifu Lin, "Supervision, Incentives, and the Optimum Size of a Labor-Managed Firm," Discussion Paper No. 525 (New Haven: Economic Growth Center, Yale University, 1987).
4. The other reforms included diversification of the rural economy, production specialization, crop selection in accordance with regional comparative advantages, expansion of local fairs, marked rises in state procurement prices, and rapid growth in the availability and better allocation of chemical fertilizer.
5. See *supra* note 2.
6. The growth rate of the agricultural component of national income dropped from the height of 14.5 percent in 1984 to 5.5 percent in 1985 and 3.5 percent in 1986. The reason that the institutional change has a one-time discrete impact on

productivity is discussed *ibid*.
7. The national income from agriculture in China grew at a rate of 3.7 percent during the first five-year plan, which started in 1953. The growth rates were -5.8 percent, 3.0 percent, 3.5 percent, 2.2 percent respectively for the second, third, fourth, and fifth five-year plans (National Statistics Bureau, *loc. cit.*) For the average growth rates of agricultural production between 1965-1980 and 1980-1985 in the developing as well as developed countries, see Table 2 in World Bank, *World Development Report 1987* (New York: Oxford University Press, 1987).
8. In addition to the disparity in marginal products created by the household responsibility system, the allocative inefficiency has another source in China. Like any other country, the endowment of land and other resources varies greatly from region to region. Since migration between rural areas failed to exist at any significant level in the past, the difference in land-person ratio as well as land-labor ratio has long been maintained. Therefore, marginal products in land and labor should also be different across regions. See the discussion in the next section.
9. My estimate is as follows: the increase in inputs and technological change can result in three to four percent growth; and the other two to three percent growth can be obtained from the improvement in resource allocation.
10. Hans P. Binswanger and Mark R. Rosenzweig, "Behavioral and Material Determinants of Production Relations in Agriculture," *Journal of Development Studies* 22, April 1986, pp. 504-539.
11. See "Several Issues Concerning the Further Strengthening and Perfecting of the Agriculture Production Responsibility System," *China Agricultural Yearbook 1981*, pp. 409-411.
12. Shi Zilu, "The Outmigration of Agricultural Labor Force in Shanghai's Suburbs Should be Given Special Attentions," *Nongye jingji wenti* 1, 1981, pp. 29-33.
13. Thomas B. Wiens, *The Microeconomics of Peasant Economy: China, 1920-1940*, (New York: Garland, Inc., 1982) p. 95.
14. Acquired ability through education will be different depending on the availability and quality of education. In this respect, the labor force in urban areas will have a higher quality than in rural areas because, in general, the education in urban areas is better and more accessible. However, within a rural area, the quality and availability of education faced by each team can be assumed

to be roughly identical.
15. Dwight Perkins and S. Yusuf, *Rural Development in China* (Baltimore: Johns Hopkins University Press, 1984), p. 83.
16. Hu Zhanyuan *et al*, "Views on Several Issues Relating to the Lengthening of Land Contracts," *Xueshu luntan* 5, 1984, pp. 71-75.
17. The government has encouraged households in a team to exchange their land in order to reduce the fragmentation of land holding. However, a household still owns more than five pieces of land, in general *China Agricultural Yearbook 1985*, p. 285.
18. Nanchong Prefectural Government Research Unit, Sichuan Province, "Scale of Land Operation and its Tendency," *Nongye jishu jinqji* May 5, 1985, pp. 28-30.
19. Selling or leasing of farm land to industries has been found to be very profitable in the suburbs of big cities (see *supra* note 12), although they are repeatedly prohibited by the government. These transactions are out of the scope of this study. This paper only looks at land transactions within agricultural sectors.
20. Fan Genxing, "A Survey of the Approaches Used in Recontracting Land," *Fujiang luntan* 7, 1984, pp. 45-46.
21. Zhou Qiren and Du Ying, "On the Specialized Household." *Zhongguo sehui kexue* 1, 1984, pp. 73-92.
22. See *supra* note 10.
23. A worker will be accepted only if his value of marginal product is greater than the value of the average product. Therefore, if any rent exists, the size of a collective team will be smaller than the size of a capitalist team because the latter only has to pay the new worker his value of marginal product. See Benjamin Ward, "The Firm in Illyria: Market Syndicalism," *American Economic Review* 48, September 1958, pp. 566-589.
24. See *supra* note 11.
25. Zhang Songmao reported in his paper that a household hired 102 workers for manufacturing electric applicants and made 200,000 *yuan* profit in 1984. See Zhang Songmao, "An Analysis of Labor-hiring in Rural Areas," *Jianghai xuekan* 5, 1985, pp. 42-44.
26. This "close one eye" policy is vividly reflected by the attitude of a party leader. Having been briefed on a trip to Hainan, Guangdong Province, that a person rented a sugar cane plantation from a collective and hired 20 workers to run it with a certain degree of success, he responded that (a) this story should not be carried in the newspapers; (b) this person's

business should not be stopped. See Zhou Dexing, "A Study of Rural Labor-hiring," *Hainan daxue xuebao* 2, 1984, pp. 25-35. There are some indications that the limitation on the number hired may be lifted soon.
27. Zhang Songmao, *op. cit.*
28. Zhou, *op. cit.*
29. See *supra* note 21.
30. See *supra* note 12.
31. Peng Xingxing and Zhang Zhaoan, "An Analysis of Migration from Other Provinces into the Suburbs of Shanghai," *Chinese Rural Economics* 10, 1985, pp. 17-18.
32. For example, only 0.6 percent of the agricultural households in Shenyangshi (one of the largest cities in China), Liaoning Province, hired workers in 1984. The hired workers consisted of 1.8 percent of the total labor force in Shenyangshi. See Gao Xuechen and Lu Guozhi, "Adhering to the Orientation of Socialist, Cooperative Economy and Actively Guiding a Healthy Development of All Forms of Economies," *Nonqye jinqji* 5, Shenyang, 1985, pp. 15-20.
33. There have been some attempts to increase the independence of credit cooperatives from the Agricultural Bank. However, a credit cooperative in a township or even in a county may be too small to overcome the problems of synchronic timing in deposit and withdrawal, as well as of covariance of defaults risk due to the covariance of yield risk in a small area. See *supra* note 10.
34. See Qin Fengge, "On the Application of Interest Rate as a Leverage in Rural Credit Market," *Shanxi caijing xueyuan xuebao* 2, 1982, pp. 50-53. The interest rates charged for loans were raised to 0.36-0.72 percent in 1981 depending on the type of loan. See *Zhongguo nongminbao*, March 8, 1981.
35. Huang Jianhui, "On Agricultural Credit Policy," *Shanxi caijing xueyuan xuebao* 2, 1982, pp. 50-53).
36. Zhao Zizhong, "On the Changes in Focus of Agricultural Loans," *Inner Mongolia caijing xueyuan xuebao* 4, 1983, pp. 72-75.
37. See "Current Situations and Problems of Agricultural Production Fund," *Nongye jingji congkan* 1, 1980, pp. 19-20.
38. Qin Fengge, *op. cit.*
39. See *supra* note 35.
40. See *supra* note 37.
41. Deposits in many credit cooperatives were unable to be cashed because of the accumulation of bad loans. See Chen Yingjian,

"The Approach Adopted by Chaoan County in Dealing with Bad Debts by Commune Members," *Xueshu yanjiu* 3, 1980, pp. 57-60.
42. All the surveys and statistics mentioned here were reported in Sun Jin, "Why Production Funds are in Shortage in Some Rural Areas," *People's Daily*, August 10, 1981.
43. See *supra* notes 37 and 34.
44. Zhang Yushu, "On Several Issues of Rural Investment." *Hunan caijing xueyuan xuebao* 1, 1980, pp. 19-24.
45. See *supra* note 38.
46. State Statistical Bureau, People's Republic of China, *Statistical Yearbook of China 1981-1984* (Beijing: Chinese Statistical Press), p. 423.
47. Lu Jianxian, "Several Issues Concerning the Reform of Credit Cooperatives," *Nonggun gungzuo tongxun* 6, 1984, pp. 38-39.
48. See *supra* note 10.
49. Mei Liangdong, "Current Situations and Issues of Rural Private Credits," *Caimao jingji* 8, 1985, pp. 51-53.
50. Zhang Zhiping, "A Survey of Rural Private Credits in Wuchang County, Heilongjiang Province," *Jingji wenti tansou* 7, 1985, pp. 409-411.
51. *Ibid*.
52. See *supra* note 49.
53. See *supra* note 13.
54. Liu Jiemin and Liang Junqian, "Widely Mobilizing Private Funds to Accelerate the Development of a Commodity Economy," *China Agricultural Yearbook 1985*, pp. 311-312.
55. For the impacts of collateral on the interest rate and on the scope of the credit market, see *supra* note 10.
56. The other plausible hypothesis is that because the available credit is sharply constrained from the supply side, lenders can charge high interest rates and choose only to lend money to high quality borrowers; as a consequence, collateral is not necessary. This hypothesis, however, is not consistent with the observations. The amount of cash in circulation rose sharply in recent years. These rises were reflected in the dramatic increase in deposits in credit cooperatives. The deposits in credit cooperatives by individual households were 7.8 billion *yuan* at the end of 1978. It rose to 32.0 billion *yuan* at the end of 1983. (See State Statistical Bureau, *Statistical Yearbook 1984*, *op. cit.*, p. 423.) However, it is estimated that the private credit in rural China was only about 15.0 billion *yuan* in 1983 (see *supra* note 49). Rural

private housing investment, nevertheless, was 21.4 billion *yuan* in the same year (*Statistical Yearbook 1984*, p. 299). These facts seem to indicate that credit markets are constrained, not because funds are not available but, because people are reluctant to lend out their money and, therefore, most available funds are used for building houses.

57. See *supra* note 54.

SEVEN

THE IMPACT OF REFORM ON THE SAVING AND INVESTMENT MECHANISM

Bruce L. Reynolds, Union College

Two tidal changes have occurred in China's countryside in the past decade. One is peculiar to China: agricultural reform or, more generally, the end of the Stalinist "forced saving" policy. The second, less widely recognized, is a phase which any developing country reaches at some point: the end of "surplus labor" in agriculture. The share of China's agricultural labor force in the total labor force has now begun to fall sharply. By the end of this century, if not before, the absolute number of agricultural workers will begin to fall. Many of the children of today's peasant farmers will spend most of their working lives in non-farm activities. These two changes are, of course, related: it was the declining importance of agriculture in the economy, among other factors, which convinced China's leaders that the time had come to jettison the Stalinist approach.

Agricultural reform has been very successful in generating rapid increases in farm output. In the eight years after 1978, real Gross Domestic Produce (GDP) doubled. This paper explores the possibility that reform contains within it, as well, the potential for drastic decline in growth.

The model which I present suggests that the "growth dividend" from reform necessarily comes early in the process, while the "growth penalty" appears only in the longer run. Thus China in the late 1980s may be entering a particularly perilous period, where the momentum for reform in general is threatened by weakness in agriculture and heavy industry.

The fundamental source of this threat, if it exists, is a shortfall of investment in rural China. With the end of forced saving in agriculture, the responsibility for saving and investment has shifted from the Chinese government to the Chinese peasant. But weak incentives to save and invest, as well as weak institutions to serve as financial intermediaries, make it unlikely that that responsibility will be met.

This paper proceeds as follows. In Part I, I construct a simple model to illustrate the extractive mechanism used in China up to 1978. The model clarifies the reasons for the mechanism's demise; these, and the impact of the policy change on agricultural growth and peasant incomes, are presented in Part II. In Part III, I explore the evidence for a savings shortfall in China. Why is it not occurring, contrary to what the model predicts? Part IV reviews other problems in China's investment pattern, and Part V surveys current policy changes designed to correct them.

A MODEL OF FORCED SAVING IN AGRICULTURE

Figure 1 illustrates the potential for trade between a rural agricultural population and an urban industrial population.

THE IMPACT OF REFORM ON THE SAVING AND INVESTMENT MECHANISM

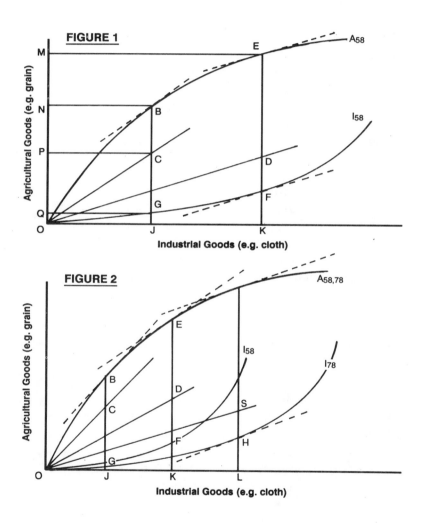

The horizontal axis measures industrial goods (in physical units)—textiles, perhaps, or bicycles, or the whole set of consumer products being manufactured in the urban sector for sale to peasants. The vertical axis measures agricultural goods—principally grain, again in physical terms—and only the grain which is above and beyond the peasants' own consumption, and available for sale to the urban sector.

The curve OA is an agricultural peasant "offer curve." Each point on OA shows, for a given quantity of cloth, the maximum amount of grain which peasant producers would willingly offer in return. Thus, for example, point B on OA shows that if OJ units of cloth were available for sale, peasant producers would be willing to trade as much as ON units of grain (although, of course, they would prefer to obtain OJ for less than ON). Similarly, Point E shows that in exchange for OK units of cloth, peasants would trade up to OM units of grain. Such a trade would leave them precisely as well off as they would be if no trade occurred at all; if they are able to obtain OK for less than OM, they would gain from the trade.

The shape of OA tells us something about the technology and preferences which underlie it. The slope of the curve diminishes as we move from left to right, indicating that peasants are decreasingly willing (at the margin) to trade grain for cloth. Two factors explain this. First, the surplus grain is being produced by applying extra peasant labor to a fixed stock of land; then the extra grain produced by each additional hour of labor must diminish. In other words, since only so much grain can be squeezed out of the land, each additional unit of grain comes at ever higher cost in terms of labor (and thus in terms of wage goods demanded). Second, in due course peasants might have bought enough cloth, and decide that leisure is a higher priority.[1]

The curve OI is the corresponding industrial offer curve. If one has in mind an urban sector consisting of numerous private cloth producers, then OI shows the grain which is needed to minimally recompense those producers for any given quantity of cloth. Alternatively, one can think of OI as showing the production possibilities of a state-owned industrial sector, which uses agricultural output to feed urban workers, to provide industrial raw materials, to trade abroad for machinery, and so forth. Point G shows that an amount OJ of cloth could be produced in return for only OQ units of grain. If more than OQ units can be obtained, the urban sector will gain from the trade. The shape of OI is similar to that of OA to some extent: the curve bows outward. As with OA, this is because the amount of the given sector's output (here, wage goods) which will be offered in response to a given quantity of the other good diminishes as trade proceeds. Both sectors become progressively less willing, or able, to trade. But for OI, as opposed to OA, the curvature is very slight. This embodies the assumption that additional capital (and labor) can be supplied to urban industry at virtually constant cost, whereas in agriculture, one factor (land) is fixed.

We need now only introduce two more concepts—the gains from trade, and the relative price of cloth in terms of grain—before putting this model to work. Suppose that urban producers do in fact produce OJ units of cloth, and peasants produce ON units of grain, and both groups come to the market. The vertical distance BJ shows the *maximum* grain peasants are *willing* to give in exchange for OJ; the vertical distance GJ shows the *minimum* amount they *must* give. Then the remaining distance, BG, is the potential gain from trade. (An artful peasant might bargain so well that the urban producers received only GJ; in that case, all the gains from trade would accrue to the peasant. But with many peasants, and many urban sellers, we predict that markets and prices, rather than bargaining skill, will determine the way in which the gains from trade are split up.)

If we are concerned not with how the grains are split up, but rather with maximizing the total gains from this trading process, how much grain and cloth should be produced? Logically, we would wish to move to the right until the vertical distance between the two curves is as great as possible. This will happen where the slope of OA equals the slope of OI. In Figure 1, this occurs at points E and F.[2]

What about the prices and markets? Consider the line OC (drawn with a slope equal to the slope of OA at B). The slope of OC is the distance CJ divided by the distance OJ. Its units are pounds of marketed grain per yard of cloth. This slope can be thought of as a price ratio: the price of cloth in terms of grain.[3] Because this price is a relative price—kilograms of grain per yard of cloth—it is usually called the "terms of trade" between agriculture and industry.

Now, suppose that this were, in fact, the price prevailing in the marketplace—that is, every peasant could exchange grain for cloth at that rate. How much grain would peasants bring onto the market? Surely, an amount ON, moving to point B on their offer curve. To the left of point B, one more yard of cloth is worth more to peasants (based on the amount of grain they are prepared to produce in exchange for it) than the price they actually have to pay in the market (i.e., the slope of OA to the left of B is steeper than the slope of OC). Peasants will make a profit on each kilogram of grain produced. Then peasants will push production of market grain to the point where this is no longer true—point B.

Of course, urban producers, making the same sort of calculation, will not settle at point G. Peasants are in equilibrium (buying and selling just what they would wish to) at point B and price OC; urban producers are not. They find the price of cloth high relative to their production cost—(the slope of OC is greater than the slope of OI at G). They will push output of textiles past OJ.

This competition by textile producers for the peasant

market will drive down the price of wage goods—the terms of trade will shift from OC toward OD. As the price of cloth falls, peasants respond by moving rightward from B. General equilibrium will be reached at points E and F where price (equal to the slope of the ray OD) equals the slope of OA, equals the slope of OI.

Thus we see that, assuming sufficient competition among both rural and urban producers, the underlying technologies and preferences, embodied in the two offer curves, determine a specific outcome: equilibrium output of OK units of grain and OM units of cloth. The total gain from trade, which is at a maximum, is the vertical distance EF. Given the market price (=slope of OD), this gain goes principally to the peasant producers: ED is clearly larger than DF.

This interesting asymmetry—peasants doing better from urban-rural trade than industrial producers—is a well-known and quite predictable phenomenon. It flows directly from the shapes of the two offer curves, which in turn reflect technological realities: peasant production of surplus grain for market is soon curtailed by scarce land, whereas urban producers have access to ample supplies of labor and capital and a replicable technology. Indeed, it is the remarkable productivity of the industrial process which will drive down the price of cloth, and give to peasants the lion's share of the gains from trade.

The application of this model to China's circumstances in the early 1950s should be immediately obvious. As the new Chinese government extended its control of the economy, by establishing supply and marketing cooperatives for sales to peasants, and a state monopoly on grain purchases, it was in effect bringing under its own control the terms of trade governing this urban-rural interaction. At the same time, the government was keenly interested in mobilizing resources for rapid industrialization. A natural way to accomplish this was to manipulate the terms of trade in such a way as to maximize, not the total gains from trade, but rather the gains accruing to the urban sector. By pushing up the price of industrial goods, the government inhibited total trade—peasant producers moved to the left from point E; but total profits to urban enterprises, plus total grain procurement, rose. In Figure 1, the higher cloth price OC illustrates this situation. Total output of grain and cloth have been reduced, from OM and OK to ON and OJ. Total gains from trade fall, from EF to BG. But gains to the urban sector increase sharply, from DF to CG. Part of this gain doubtless was absorbed by high industrial wages. But the shift must also account in part for the rise in the Chinese rate of saving and investment, from below 10 percent in 1950 to well above 20 percent in 1957.[4]

The extent of this "forced extraction" has been a matter of considerable debate. Some authors (including many Chinese economists) have even argued that the terms of trade shifted in favor of agriculture in the period prior to 1978, rather than the other way around. But the single most thorough examination of this issue concludes[5] that "enough evidence exists to postulate that the state transferred significant resources out of the agricultural sector" in the 1956-1978 period. On the presumption, then, that the policy was pursued, and was effective: why was it abandoned in 1978? Part II addresses this question, and charts the shift in strategy in the past 10 years.

1979: THE END OF FORCED SAVING IN AGRICULTURE

A policy gives way when either the costs of the policy have greatly multiplied, or the benefits have diminished. The Stalinist approach to agriculture was abandoned for both reasons. Figure 2 shows one part of the rising costs.

In the two decades after 1958, Chinese productive

capacity expanded substantially, both in agriculture and in industry. But since the expansion was more pronounced in industry than agriculture, Figure 2 represents only that shift: an industrial sector which can produce wage goods at even lower cost (relative to grain) in 1978 than 20 years earlier. The outward shift in the industrial offer curve reflects the substantial increase in China's industrial plant and equipment in this period. Where before, equilibrium was at E and F and price OD, now a free market policy will take us to R and H and a (lower) cloth price shown by the slope of the ray OS [equal to the slope of OA at R = slope of OI at H].

Note that the impact of 20 years of industrial growth on the urban gains from trade under an extractive policy is nearly zero. Extraction is only marginally greater than CG; industrialization has no pay-off here. Meanwhile, the policy's constraining effect on output is ever more costly. A return to market allocation, which in 1958 might have doubled industrial goods output (OJ vs. OK), now will triple it (OJ vs. OL). The cost in terms of total gains from trade is also rising (RH is larger than EF).

The benefit of maintaining this extractive approach was also less by 1978 in a relative sense. As Table 1 shows, in an increasingly industrialized China, agriculture's ability to affect the rest of the economy was diminishing. Its share of the labor force had fallen sharply. Its share of the value of output was also smaller: 34 percent in 1985 as against 57 percent in 1952. (If the value of services were included, the drop would be even more pronounced.)

Beginning in 1979, the extractive policy was rapidly dismantled. By 1983, purchase prices for farm and sideline products had risen by 48 percent compared with 1978.[6] For the period 1978-1985, agricultural purchase prices rose 67 percent *versus* a rise in consumer goods of 28 percent, for a terms of trade gain of 29 percent.[8] In addition, the scope of

Table 1
AGRICULTURE'S ROLE IN THE ECONOMY[7]

	Rural Population (% of Total)	GVAO (% of GVIAO)
1952	88	57
1957	85	43
1965	82	38
1978	82	28
1982	78	34
1985	63	34

mandatory sales of grain and other products to the state has been reduced. This system, put into place in January 1961, required sale to the state at state-set prices for a certain portion of the crop of almost every agricultural product. By the end of 1984, coverage had been reduced to 14 major crops, from 25 crops in 1961.[9] Lastly, price-setting by the state has slowly begun to give way to free market prices. There are three categories of goods for price-setting purposes: Category I (price set by the state), Category II (price set by the central government, but local governments have power to vary the price seasonally), and Category III (price set by market forces). In late 1986, the distribution of agricultural products among these three categories (in terms of value of sales) was 37 percent, 23 percent and 40 percent respectively; the policy intent was to gradually increase the portion marketed in category III.[10] By early 1987, only 25 kinds of agricultural products and byproducts had their prices set by the state.[11]

This shift in the terms of trade was reinforced by the production responsibility system, which ensured that the benefit of higher prices accrued directly to peasant farmers who responded with higher output. The cumulative effect of this new policy is clear from Table 2. Sales of wage goods to peasants more than tripled between 1978 and 1985. Sales of farm and sideline produces also tripled and grain purchases more than doubled. (The value measures in Table 2 should be discounted for a general price rise of perhaps 30 percent over this period.)

Table 2
INTERSECTORAL SALES
(billion *yuan*)[12]

	1978	1983	1985
Rural Sales of Consumer Goods	81	148	252
Peasant Sales of Farm and Sideline Products	56	126	169
of which:			
Direct Sales to Non-agricultural Residents	3	11	29
Sales to Commercial Departments	46	99*	107
Sales to Industrial or other Departments	7	14*	33
Volume of Grain Purchases (million tons)	51	117*	108

*1984

The abandonment of a Stalinist approach to agriculture, then, has had a major effect on agricultural output, just as Figures 1 and 2 would suggest. The other implication of Figure 1 is that this new policy would sharply reduce the ability of the central government to capture the lion's share of the gains from trade, and thus reduce its ability to finance investment activity. This other effect is also dramatically obvious in the years since 1978. The share of accumulation (investment) in national income fell from 36.5 percent in 1978, to 33.4 percent in 1985.[13] But more significantly, as Table 3 shows, the majority of this saving is now being done, not by government, but by enterprises and households.

Not only has the government in China ceded a significant part of responsibility for saving to enterprises and households. Investment decisions have also been dramatically decentralized. Even the reduced portion of investment which now flows through government is

Table 3
DISTRIBUTION OF GROSS DOMESTIC SAVINGS(%)[14]

	China 1978	China 1981	China 1985	Japan	South Korea	India
Government	73	49	38	9	26	11
Enterprises	12	22	25	37	35	22
Households	15	29	37	54	38	65
Total	100	100	100	100	100	100
Aggregate Savings Rate	37	29	33	32	25	22

increasingly in the form of extra-budgetary, locally-controlled funds. Local unbudgeted funds were 167 billion *yuan* in 1986, a nearly five-fold increase over 1978. In the first three months of 1987, unbudgeted local capital construction doubled compared with 1986, accounting for a large share of the overall 35 percent increase in local capital construction.

Thus the liberalized agricultural policies since 1978 have brought a mixed blessing. On the one hand, they have had an immediate, obvious and dramatic impact on output. Much of China's economic boom, since 1978, can be traced to the stimulus of startlingly high agricultural growth rates. But on the other hand, the saving and investment function—a crucial determinant of growth in the long run—has also been sharply altered. Part III explores this more worrisome aspect of the impact of rural reform on growth in China.

REFORM AND AGGREGATE INVESTMENT

Economic reform is usually analyzed (and generally applauded) by economists using the framework and the criterion of static efficiency.[15] A shift toward the market mechanism, goes the argument, will generate price signals which are accurate indicators of where resources should flow; this will mean more efficient allocation of resources. And the same mechanism, by linking profitability to rewards, will cause producers in every sector of the economy to conserve on resources—to produce efficiently. These two desiderata—allocative efficiency and production efficiency—are the areas in which market economies excel. They squeeze the most possible output from the resources currently available.

The primary goal of planned economies, however has usually been, not static efficiency, but rapid and sustained economic growth. It was at the altar of the Growth God, not the Efficiency God, that strong central planning was sacrificed in China in 1979. Unless reform guarantees growth, its future is dim.

The model embodied in Figure 2 shows a growth impact from rural reform: the increased output of grain and cloth as the economy transits from points B and G to R and H. The doubling of Chinese GDP in the eight years 1979-1986—an annual growth rate of 10.1 percent—amply bears this prediction out.[16] But this growth largely is the one-shot result of static efficiency gains. It reflects the reallocation of labor out of farming, and into rural small-scale industry—70 million workers in the period 1978-1986. [17] It also stems from enormously increased efficiency in the use of the remaining agricultural labor force—the production responsibility system causing peasants to use their own labor well.

The major source of growth in China over the last 30 years has not been rising static efficiency, but rather a high rate of saving and investment, generated by and channeled through government. The second prediction of Figures 1 and 2 is that this flow of investable funds is sharply diminished by reform. The total gains from trade rise, but now the lion's share goes, not to the urban, state-owned industrial sector, but to the rural sector. Will peasants save and invest 100 percent of this windfall? If not, will aggregate saving and investment fall? And, if so, is it possible that when the dust has cleared from the initial impact of reform on static efficiency, China may find her growth rate dropping sharply? In the rest of this section, I explore the impact of reform on the aggregate investment rate. Part IV discusses the impact of reform on the efficiency of investment, and Part V outlines policy changes which may mitigate some of the problems presented here.

The economic facts of the past eight years confirm the story told by Figure 2. Per capita rural incomes rose from 134 *yuan* in 1978, to 424 *yuan* in 1986.[18] In real terms, per capita rural income rose 14.8 percent per year from 1978 to 1985.[19] A corresponding decrease in government control over saving and investment is also clear. As the figures in Table 3 show, between 1978 and 1981, government saving constituted a decreasing share (73 percent down to 49 percent) of an aggregate saving rate which was itself decreasing (from 37 percent down to 29 percent of GDP). By 1981, the Chinese government, which three years earlier had control over a flow of saving constituting 27 percent of GDP, controlled only 14 percent. This trend continued from 1981 to 1985. If, instead, we look at investment, and ask: what proportion of national income was devoted to capital construction through state-owned units? The picture is the same: the figure falls from 15.1 percent in 1978, to 8.5 percent in 1985.[20]

And yet, despite this apparent collapse of government control over the saving and investment process, China's investment rate did not decline markedly in this period. The ratio of accumulation to national income in 1985 was 33.7 percent, not sharply different from 1978's 36.5 percent. The question arises: if government is not doing the saving, who is?

Rural Chinese are clearly doing more saving than in the past. Individual bank savings deposits rose from 21 to 230 billion *yuan* in this eight-year period.[21] Peasants are also investing, but "unproductively"—in housing, consumer durables, and goodwill (lavish weddings).[22] But peasant saving and direct peasant investment could not account for the continued high aggregate saving and investment rate in China. The answer to the puzzle is this: not all of the "lost" gains from trade have reached peasant pockets. Important industrial products—notably fuel and fertilizer—are still central government monopolies. Other agricultural inputs, and manufactured consumer goods, are increasingly produced by local-government enterprises, or by township and village enterprises; but here too, price is still higher than the "competitive equilibrium" price of Figure 2, because these government authorities use administrative power to maintain a local monopoly, and keep price and profit high.

Evidence abounds of the monopoly profits being reaped by Chinese industry. Between 1979 and 1983, township and village enterprises (TVE)—a minor part of China's industrial sector—earned 56.9 billion *yuan* in profits. Of this, 18.6 billion *yuan* went to the state in taxes, and another 9.4 billion *yuan* was "poured directly into agricultural production—buying machinery, improving the land, and sharing their income with members of their communities who continue to farm."[23] Profits from the TVE sector contributed six billion *yuan* to agricultural production in 1981-1985 (comparable to total state investment in rural capital construction!), and also raised 23.2 billion *yuan* building marketplaces and for rural health, education and welfare expenditures.[24]

It is clear that these enterprises, in the relaxed environment since 1978, have in effect spliced themselves into the extractive loop which was previously an exclusive central government preserve. As one survey of 200 large-scale TVEs puts it: "Owing to the wide disparity in price between industrial and agricultural products, these enterprises have made high profits."[25] To some extent, the central government recaptures this flow, through tax and profit remission. To some extent, local government captures a share, through the imposition of ad hoc taxes (generating "extra-budgetary funds"). And to some extent, the profits simply pile up within these medium- and small-scale enterprises.

The first effect of this new state of affairs—industrial producers which still have monopoly power *vis-a-vis* the rural sector, but which are decreasingly under central government control—is a new division of the gains from trade. Compared with the division depicted by Figure 2's "competitive market equilibrium" (RS to peasants, SH to industry), grain producers are evidently getting a somewhat smaller slice of the pie; the workers and "owners" of these locally controlled industrial enterprises are getting a slice as well. The second effect of the new situation is to keep investment rates relatively high—higher than a "pure Figure 2" result would have done. After all, the "owners" of China's rapidly-expanding local industrial sector (the hundreds of thousands of enterprises which are not directly controlled by Beijing) do not really own these assets. They are inhibited by socialized ownership (and by regulations) from passing all the profits along to individuals (who might simply consume them). No such inhibition exists against using profits to expand production. On the contrary, "Expand production!" has been the watch cry of nearly four decades of central planning in industry. So the gains from trade, although no longer controlled by the central government bureaucracy, still end up being invested.

But the investment decisions are being made by different people from before—by enterprise managers, county party secretaries, city mayors, and the like, instead of the State Planning Commission and the State Council. Therefore, the pattern of investment will necessarily be different after 1978, compared with before. This change may also threaten continued growth. (Indeed, some of China's most ardent reformers argue fiercely that this is the case.[26]) It is to this issue—the danger of inefficient allocation of investment—that we turn in Part IV.

REFORM AND THE EFFICIENCY OF INVESTMENT

There are three different ways in which reform may have affected the efficiency with which investment funds are now used in China: sectoral allocation, scale, and project selection. It may be that investment is now going to the "wrong" sectors; the widespread claim is that too much is going to light industry (and to "nonproductive" sectors), and too little to agriculture and heavy industry. Or it might be that within each of these sectors, the projects being financed are not those with the highest rate of return. Lastly, it may be that because the units making the investment decisions are at a lower level, with each having only its own, smaller pool of investment funds to draw on (and its own regional interests in mind), new plants are too small in scale to reap the benefits of mass production techniques. We will address each of these points in turn.

Sectoral Allocation

Table 4 shows that there have indeed been shifts since 1978 in the pattern of capital construction investment by state-owned units. Both industry and agriculture have suffered. A breakdown of industry by sector would show that heavy industry has suffered, while light industry has gained. Transport, energy and raw materials, for example, have seen their pre-1979 50 percent share of investment funds fall to 33 percent in 1985.[27]

Urban infrastructure, the commercial network, and science and HEW have picked up the slack. This pattern conforms to what one would expect, given the fact that local governments now control a large part of the investment flow. An increasing part of this capital construction budget (20 percent in 1978, 40 percent in 1985) is financed through extra-budgetary funds—funds which are "self-collected" by local governments and enterprises, through special taxes, retained profits and depreciation funds.[29]

It is clear that this shift in the sectoral pattern of investment goes against the preferences of the central

Table 4
SECTORAL ALLOCATION OF CAPITAL CONSTRUCTION
FUNDS BY STATE-OWNED ENTERPRISES
(percent)

	1978	1981	1985
Industry	54.5	48.8	42.0
Agriculture	10.6	6.6	3.4
Urban Public Utilities	3.1	7.21	8.6
Commerce, Materials Supply	3.1	6.3	6.7
Science	4.3	9.8	11.2

government. For example, Zhou Daojiang, President of the People's Construction Bank of China, reports "unauthorized investment" of 20 billion *yuan* in 1986, and warns that the 160,000 current investment projects is excessive. "When local governments and enterprises have money," he notes, "they invest as they like, and you can hardly do anything about them." Zhou complains that "the State should not have allowed local governments and enterprises to keep so much money."[30] But can we conclude from these sorts of jeremiads that the present pattern is less efficient than the pattern which the center would impose?

On the contrary: a number of the sharp shifts which we have witnessed in the investment allocation in the past eight years seem quite rational. When China's past pattern of production is thrown in relief against the experience of low-income market economies, several "aberrations" stand out:

- urban housing per capita is low.
- industry is "muscle-bound": a disproportionately large capacity in heavy industry (producer's goods) compared with light industry (consumer's goods).
- rural industry is primarily backward-linked to agriculture—that is, it features prominently the production of inputs to agriculture (machinery, fertilizer) but slights forward-linked industries (food processing, transportation, storage, and provision of consumer goods and services).

There is a specific and useful methodology for testing the rationality of sectoral investment allocation.[31] The calculations for such a test cannot be undertaken here. But impressionistically, I see no compelling case that the present pattern of investment allocation is irrational.

The exception to this conclusion is agriculture. Even here, it is difficult to draw firm conclusions, since most of the flow of investment to agriculture is not captured by national statistics. But several things can be concluded (see Table 5). First, state investment in agriculture is minuscule. Second, the other main source of state investment funds, the Agricultural Bank of China, provides an equally small amount. The main sources of investment funds are the resources of collective industry and collective agricultural units themselves.

The size of individual self-investment in agriculture by farmers themselves is uncertain, but it is probably negligible. Farmers, after all, have at most a 15-year lease on the land in which they might invest, and for many, half or more of those years have already run. A recent article in *Farmer's Daily*, for example, cited reports of a precipitous decline in productive expenditures by farmers, and explained, "Some farmers fear a change in the current policy, saying, 'I'd rather spend the money on myself than throw it in the grain fields.'"[33]

Table 5
INVESTMENT IN AGRICULTURE, 1985
(billion *yuan*)[32]

Capital Construction by state-owned units	4.8
Increase in Rural loans of the state banking system	5.1
Rural Collective enterprise profits allocated to agricultural investment	17.1
Collective self-employment	13.0
Individual self-employment	n.a.
	40.0

Even a crude application of the rationality test in footnote 31 demonstrates that 40 billion *yuan* is too little to invest in agriculture, compared with investment in industry. Investment in industry cannot be less than 160 billion *yuan*—four times as large. Even if the desired growth rate in industry were three times that in agriculture (12 percent *versus* 4 percent), and even taking into account the larger gross output of the industrial sector, the irrationality is apparent.[34]

The problem of agricultural investment is unmistakable. A symposium on problems of rural reform in March 1987, underscored the dangers of the present trends. Cultivated land has fallen by 2.67 million hectares since 1978. State investment in irrigation construction dropped by half between 1980, and 1985, to 1.1 billion *yuan*. This, plus poor maintenance, has caused irrigated acreage to decline by nearly a million hectares over this period. Mechanically farmed acreage has declined by six million hectares, and the agricultural technology extension system is described as "in tatters."[35] A great deal of the peasant momentum in favor of reform grew out of the surge in agricultural production up to 1984. The stagnation of output since then has increasingly undercut the credibility of overall reform.

THE IMPACT OF REFORM ON THE SAVING AND INVESTMENT MECHANISM

Scale and Project Selection

Because the investing unit in China is now typically at a low hierarchical level (a local government, or an enterprise), a bias has been introduced toward small investment scale. Across the whole range of consumer manufactures (light industry, with its artificially high profit rates, is heavily favored), this trend appears: in tobacco, textiles, electrical appliances and the like. The bias arises in part because a given investing authority may not have enough funds to build a large plant, and no financial intermediaries exist to permit lending and borrowing across administrative boundaries. But even if ample funds were available, each administrative unit—province, prefecture, county or town—is likely to build a plant with just its own market in mind, rather than building a larger plant and selling its output nationally.

The reason is simple: each provincial and even sub-provincial authority in China has the power to erect trade barriers, excluding competition and preserving monopoly control over its own market. Hence the startling replication of production facilities across the country: 112 assembly lines for color TV sets, for example. In complaining of the misuse of TV investment funds, an *Economic Daily* article complains that they "invest in unnecessary projects, setting up highly profitable industries at the expense of low-profit industries which are nonetheless vital to a sustained agricultural growth....Townships will invest only within their own administrative region."[36]

To summarize: given the dramatic restructuring of China's system for mobilizing savings and making investment decisions, there is ample reason to suppose that the pattern of investment allocation might have shifted sharply—and indeed, we see that this is the case. With the glaring exception of agriculture—which has always been underfunded—there is no clear evidence that the sectoral distribution of investment is less rational now than prior to 1978. Concerning scale of project, there is ample *a priori* reason, and considerable anecdotal evidence, to suppose that the scale of investment in China today is irrationally small. The concluding section discusses policy changes which would mitigate both of these problems.

CONSTRAINING INVESTMENT BEHAVIOR: MORE REFORM, OR LESS?

The policy recommendations which surfaced in the first half of 1987, designed to correct the problems described above, fall into two groups. Some changes would re-centralize; others would push reform further, with an eye to making the decentralized system work more effectively.

Among attempts to bolster the role of government in the investment process, to correct sectoral imbalance in investment, some are quite direct and heavy-handed. In Zhejiang Province, for example, concern with the "drastic fall in agricultural investment, neglect of irrigation, water conservation systems and farm capital construction" led to the institution of a direct labor tax: each farmer must work 10 to 15 days per year on these projects. In addition, the provincial government complained that township and village enterprises "turn part of their profits into cash dividends for farmers.... But instead of investing money in agricultural production, the farmers spend it on consumer goods." So the provincial government "has stipulated that such profits be turned into 'farm development funds,' " and reports that "The farmers welcome(!) the practice of changing dividends into a free supply of (improved) seeds."[37] Jiangsu instituted similar measures in 1986, requiring five million farmers to work on irrigation projects, and exacting contributions for agricultural development from all Jiangsu enterprises. These amounted to 100 million *yuan*, much larger than government agricultural investment.[38]

Other forms of re-centralization are more indirect: the imposition of taxes, or soaking up funds by issuing special purpose bonds. The State Council, for example, imposed in May 1987, a special seven percent tax on the profits of all non-state enterprises, the proceeds of which are earmarked for investment in energy and communications. This is an attempt to redirect some part of local extra-budgetary funds, which swelled to 167 billion *yuan* in 1986.[39] Lastly, the Agricultural Bank of China has encouraged inter-branch lending and borrowing, and is selling bonds in cities earmarked for agricultural development. The amounts involved are still small, but the idea makes sense.

As opposed to these relatively predictable responses to the problems seen in Part IV, some responses put forward in the first half of 1987 amount to making reform work better by pushing it further. To correct the problem of farmers failing to invest in agriculture, these proposals call for the extension of land ownership rights. *Farmers Daily* in April called for an assurance of policy continuity, the power to transfer land to more productive farmers, and an acceptance that "private property must not be violated except for taxes."[40] And to solve the difficulty of small investment scale, reformers are pushing for the development of inter-regional money markets—institutionalized, continuous arrangements for interbank and inter-enterprise lending and borrowing.

CONCLUSION

This paper has presented a model designed to clarify the nature of the changes in China's agricultural system since 1978. The model explains the marked growth in agricultural and light industry output since 1978, and suggests that this growth spurt is in part a one-shot phenomenon due to static efficiency gains. It also predicts a sharp redirection of the gains from inter-sectoral trade—a loss of the government's ability to mobilize investable funds—which, if it produced a drop in the investment rate, would be quite threatening to future growth in China.

Interestingly, the fall in the investment rate which the model predicts has not occurred. As explained in Part III, while the central government did lose power over investment flows, local governments and local enterprises captured much of the flow which the center had lost, and hastened to invest the funds themselves. The pattern of that investment, however, differed markedly from central government preference. Part IV explored the implications of that divergence, and suggested that here, too, no serious problems appear (aside from underinvestment in agriculture, a long-standing phenomenon in China). Part V reviewed current policy proposals. Some of these are reformist, market-oriented; others are more heavy-handed. But on balance, the mix seems predictable and rational.

NOTES

1. OA can be interpreted as a total product of labor curve, coupled with a backward-bending labor supply curve.
2. The slopes of the two curves are indicated by the straight lines which are drawn tangent at E and F. To the left of E and F, the slope of OA exceeds that of OI; the curves are drawing farther and farther apart as we move from left to right. But since OA's slope is diminishing, and OI's is increasing, the rate at which the distance between the two curves increases must diminish, until (at F and E) it is zero. Here, the distance between the curves is greater. Thereafter, they converge.
3. Suppose, for example, that grain sells for two *yuan* per kilogram, and cloth for one *yuan* per yard. Then the relative price of cloth is 1/2: it costs 1/2 kilogram (one *yuan*'s worth) of grain to obtain one yard of cloth. The slope of OC in this case would be 1/2.
4. Planners could move peasant producers past point B, and increase the total surplus extracted, by using price discrimination: a higher grain price (lower cloth price) for grain sales above ON. And, in fact, the Chinese did use such a system: the "compulsory sales price," the higher "above-quota price," and the even higher "negotiated price."
5. Nicholas R. Lardy, *Agriculture in China's Modern Economic Development* (New York: Cambridge University Press, 1983), p. 127.
6. *China Agricultural Yearbook, 1985*, p. 33.
7. State Statistical Bureau, *People's Republic of China, China: A Statistical Survey in 1986* (Beijing: New World Press, 1986), pp. 9, 22.
8. State Statistical Bureau, *op. cit.*, p. 100.
9. See *supra* note 7.
10. Frederick Crook, Trip Notes (unpublished, 1986).
11. *China Daily*, June 6, 1987, p. 1.
12. State Statistical Bureau, *op. cit.*, pp. 86-94.
13. *Ibid.*, p. 6.
14. World Bank, *China: Long-Term Development Issues and Options* (Baltimore: The Johns Hopkins Press, 1985), p. 145. The figures for 1985 are a crude estimate using data from the State Statistical Bureau, and embodying the following assumptions: (1) total investment equals total savings; (2) of the fixed investment by

state-owned units which is not capital construction, 10 percent is from extra-budgetary funds; (3) all extra-budgetary funds are enterprise saving; (4) all urban saving deposits are individual; (5) all household saving which is not directly invested by individuals is invested by collective or state enterprises.
15. For examples of this approach see William Byrd, "The Impact of the Two-Tier Plan/Market System in Chinese Industry," in Bruce Reynolds, ed., *Chinese Economic Reform* (Cambridge Massachusetts: Harcourt, forthcoming), 1987, and Justin Yifu Lin, "Rural Factor Markets in China after the Household Responsibility System Reform," presented to this conference (Manila, August 1987).
16. *China Daily*, June 4, 1987, p. 1.
17. *China Daily*, April 16, 1987, p. 4.
18. See *supra* note 16.
19. *Beijing Review* 19, May 11, 1987, p. 14.
20. State Statistical Bureau, *op. cit.*, pp. 83, 84.
21. See *supra* note 16.
22. Per capita peasant spending on housing rose from 3.7 to 39.5 *yuan*, 1978 to 1985. Three in every 1000 farm households will buy a motor vehicle in 1987. A State Statistical Bureau survey in 1986 found that the average expenditure on weddings in three coastal provinces was 4,000 *yuan*. State Statistical Bureau, *op. cit.*, p. 111, and *ibid.*, p. 2.
23. *China Daily*, November 15, 1986, p. 4.
24. *Beijing Review* 23, June 8, 1987, p. 20.
25. *Ibid.*, p. 22.
26. See Bruce L. Reynolds, ed., *Economic Reform in China: Challenges and Choices* (New York: M.E. Sharpe Inc., 1987).
27. *China Daily*, March 9, 1987, p. 4.
28. State Statistical Bureau, *op. cit.*, pp. 71, 73.
29. See *supra* note 27.
30. *China Daily*, March 18, 1987, p. 1.
31. Essentially, the appropriate test, for any two sectors (say, light industry *versus* heavy industry) is: does the ratio of the investment in each of the two sectors equal the ratio of their desired growth rates (weighted by their shares of total output and by the inverse of the ratio of the incremental capital-output ratios of the two sectors)?
32. State Statistical Bureau, *op. cit., passim*.
33. See *supra* note 27.

34. The appropriate test shows: the ratio of investment in industry to investment in agriculture (160/40 or 4) is far larger than the ratio of the desired growth rates (3?) times the sectoral shares (560/220 billion *yuan* or 2.5) times the inverse of the ratios of the sectoral incremental capital-output ratios (1.5 for industry, 0.3 for croplands; inverse is 1/5 or 0.20). 3x3.5x.2=1.5, far less than 4. Data on sectoral shares and IKORs from World Bank, *op. cit.*, Annex 4, Table 4.1 and Annex 5, Table C4 respectively.
35. *China Daily*, March 30, 1987, p. 4.
36. *China Daily*, February 3, 1987, p. 4.
37. *China Daily*, May 28, 1987, p. 4.
38. *China Daily*, March 25, 1987, p. 3.
39. *Beijing Review* 19, *op. cit.*, p. 4.
40. *China Daily*, April 6, 1987, p. 4.

EIGHT

CITY, PROVINCE AND REGION
THE CASE OF WUHAN

Dorothy J. Solinger, University of California–Irvine

> ...regarding the positive role that the old economic system played in the country's socialist construction in the past, its positive aspects, and our positive experience in our past work, we must not adopt a nihilist attitude and negate everything. Thus, in designing target patterns for reforms, we are required to take into consideration the problem of replacing one pattern by another, the need to keep the economy running, and the problem of continuity between patterns.[1]

This quotation from a *People's Daily* article by Liu Guoguang, one of the country's leading theoretical economists, captures very succinctly what has been the—often unintended—experience in Wuhan's economic reform. Right at the heart of the process of economic reform lies the issue of replacing that "old economic" system's vertical, bureaucratically organized channels of command and coordination with lateral, market-style linkages.

The idea behind this move is that, given some freedom, lower-level entities should choose horizontal bonds, since these are more economically rational than the old, planned ones. But to what extent can a command to reform, to alter old economic patterns into drastically different designs, really be obeyed in a short stretch of time? If, instead, there are crucial continuities, just what form do they take?

This paper, based in part on newspaper and journal articles from the past three years, but mainly on 35 hours of interviews conducted in Wuhan in June 1987, addresses this topic. In addition to the big question just posed, it raises these correlative queries as well: Is decentralizing and reorienting reform really occurring in and around Wuhan and, if so, is it empowering the city in important ways? Is a rearrangement of responsibilities, in fact, breaking down vertical, administrative, bureaucratic patterns of investment, management, and exchange? Are these new relationships then allowing, and inducing, economic agents freely to opt for horizontally-based connections?

The usual line of argumentation in analyzing the recent Chinese reform process has gone like this, for the most part: reform is blocked by bureaucratic interference, whether by "conservative" politicians and commissions in the capital, or by local departments and their cadres. For, so commentators hold, such forces fear the loss of their funds, power, and ultimate control over economic and financial activity. Such descriptions pit plan against market.

Actually, the picture is a lot more complex. For those in charge of reform have been able, in some cases, to turn the old structure to advantage. By a complicated combination of removing, yet also relying upon planks from the plan, reformers can do a great deal to bring about changes that appear to be market-based, in the spirit of the state's propaganda on reform. But, rather than junking the plan altogether, reform has in many cases been possible simply through adaptation of old plan offices, channels, personnel, and relationships.

There is a second way in which the previous system facilitates reform. Those promoting and implementing reform are sometimes doing so just by expanding, legalizing, and formalizing; in short, by making explicit the formerly semi-licit, and even illicit, coping strategies used for years whenever the old plan had proved inadequate.

And, third, reform-minded powerholders in Beijing (as well as those in cities such as Wuhan—probably through the connections such city leaders manage to fashion with reformers at the top) can, in many cases, throw the considerable weight of the state behind local models or test points (*shidian*) in order to promote particular projects they sanction. At the urban level, politicians may help models succeed in order to win a name for their city as a pioneer in reform. Political facilitation, called *youhui tiaojian* (preferential conditions) in the more economically-slanted jargon of today, thus pushes along pet reform designs. At the same time it provides a model of what the reformers hope to realize on a much larger scale in the future, when the political climate has been more fully prepared.

These three state-sponsored reform strategies at the policy level—remolding old organs and their connections for new purposes, bringing coping modes out of the shadows, and providing political facilitation for models—when put into play, evoke further efforts in the direction of reform from the cities and enterprises that benefit from them. The result is a still inchoate, but potentially developing set of new relationships between five key layers of the central-local economic hierarchy: center, province, city, bureau, and enterprise.

Before we can ask to what extent this process is leading to the substitution of vertical chains by horizontal rings, we need to take note of another trend, often in opposition to the reforming one. That one is also state-centered, as it seeks to place an overlay of state management atop the burgeoning business brought forth by the facilitation and encouragement described above.

As always, with liberalization in the economic realm in

China, state control and management offices are springing up in tandem, in an effort to ensure that altered modes of transaction do not lead to economic instability, transgress state mores, or result in the shortchanging of prime state construction projects.

The two trends come together at times as well. There can be a thin line between promotion and prevention: when commissions composed to encourage the new relationships try to channel their initiatives, or when planning bureaucracies seek to compete with the new forms or control the commodities and funds flowing through them. The effect of this is that a double image is emerging. There are flourishing freewheeling ventures, not just protected but even promoted and coddled.[2] But those ventures play to an accompaniment of state-sponsored supervision.

This is not to say that reform is being seriously stymied. In fact, many changes have occurred, or are occurring, within the state-run bureaucratic economy. To a large extent, they must be understood, not as the birth of purely free markets in China, but, on the positive side, within a context of adaptation and remolding within the bureaucracy and against a backdrop of specific political support; on the negative side, at times against a countervailing regimen of restraints.

More concretely, at the lower echelons, the reform movement appears in a mixture of modes. Cities such as Wuhan have received some relief from their provincial overseers, as have enterprises from their management departments (bureaus). But in finding their way in the new, unfamiliar "market-place," firms and municipalities in many cases draw on the very supply sources and sales outlets that were once written into their plans, rather than seeking some reorientation according to the new, supposedly economically-rational, principle of association.

Another form of this search for certainty and security is

localism, whereby factories and stores pick exchange and investment partners close at hand. Similarly, "departmentalism" crops up in a new guise as firms in the same area and trade choose to associate in what are mergers or even oligopolies to boost local economies and to rationalize the product structures of municipal or provincial trades for better competitive advantage.

The remainder of this paper will first say something about Wuhan and its significance in the urban reform experiments. It will then review four of the most progressive types of urban reforms underway in Wuhan (many of them being pilot-tested there) from 1984-1987 (the period during which Wuhan has had the status of a trial point for comprehensive urban reform). In doing so, it will assess the contribution of these reforms to an altered pattern of relationships between state, province, city, bureau, and firm.

These four reforms are: Wuhan's designation in June 1984, as a separate line-item in the state plan (*jihua danlie*) and the concomitant decentralization (*xiafang*) of some 50 central and provincially-managed enterprises to the city from autumn 1984; the schemes to separate management from ownership of firms through such measures as bankruptcy, leasing, and manager responsibility systems; new investment sources—stocks, bonds, share systems, and short-term capital markets and networks; and new circulation methods which are to shift the vertical supply and sales channels in use before reform to ones based on lateral exchange. Also in the fourth category are: enterprise groups, bidding for component parts, markets for the means of production, and trade centers. The conclusion will comment on the connection between market and plan and the functions and dysfunctions of continuity between systems in this transitional era.

WUHAN AS A PILOT CITY FOR REFORM

Wuhan was selected to pioneer an experiment in granting provincial capitals separate line-item status in the central plan in June 1984.[3] By then a year had already passed since Wuhan University Professor Li Chong-huai coined the phrase *"liang tong"* (the two *tong*'s, *liutong* and *jiaotong*, or, in English, the "two C's"—circulation and communications) to serve as the focus for this city's reform.[4]

This slogan was meant to express the particular strengths of the city, which sits astride both the Yangtze River and the Guangzhou-Beijing Railway, almost equidistant from Chongqing, Shanghai, Guangzhou, and the national capital. Both historically and under the People's Republic it has had a place of special importance as a transport hub and a transshipment center. But its traditional role as the "thoroughfare of nine provinces" was to a large degree thwarted under the rule of vertical bureaucracies, with their planned, compulsorily-conducted commerce and their transport coordinated by separate, enclosed, political authorities along administratively-demarcated boundaries.[5]

The city's mission, once it was singled out as an experimental point for "comprehensive urban reform" was to use its new powers, supposedly equal to those of a province in economic affairs, to form the core of an economic region in Central China. It could then go on to stimulate commodity production in all the surrounding hinterland and in nearby cities.[6] Moreover, it was hoped, with its new independent powers, Wuhan should be able to reenact its historical role as a center of foreign trade, finance, banking, and information, as it reinforced its newer, post-1949 position as a base of industry and of higher education and research.[7]

All in all, the purpose of the *jihua danlie* initiative was "to enhance Wuhan's economic position in the country and bring into more extensive play its role in establishing economic ties throughout the province and the nation."[8] In short, Wuhan's new powers would underline the overall national economic reform's dramatic shift in emphasis, toward commerce and economically-based, decentralized decision-making.

Under Wu Guanzheng, mayor of the city from March 1983 until fall 1986, the city became noted for "opening wide the gates" of its three joined cities (Hankow, Wuchang, and Hanyang) to outside competition. Its progress even caught the attention—and the praise—of Premier Zhao Ziyang, who announced in 1984 that he had "placed his hopes in Wuhan," and came to see for himself in April 1985, when Wuhan was "marching at the head of the urban reform."[9]

Immediately upon gaining its new status for their city, municipal leaders set about organizing trade centers, forming a wholesale network, introducing bidding on a nationwide basis for parts in light industrial production (instead of relying solely on allocations from local factories), inviting peasants from the countryside and residents of other cities to open shops and manage factories, and relaxing price controls and rationing, all of this before most—and, in the case of some of these measures, before any—other cities in the country. Other steps were taken to improve transport management and to revive the financial sector.

Now, three years after that heady first taste of autonomy, Wuhan continues to stand in the forefront of the urban reforms, being picked, for instance, as one of the pilot cities for financial reform in early 1986[10] and chosen as the site for one of seven national steel markets in the beginning of 1987. But to what extent have all of these reforms really led to a divorce of Wuhan from the vertical bonds that bound its business, and in what sense is this leading to the creation of a presumably more rational routing of economic relationships?

FOUR KINDS OF REFORMS

1. Jihua Danlie

Financial and Behavioral Outcomes

According to one Wuhan scholar, the *jihua danlie* reform must be viewed as an interlinked set of three component parts: getting separate standing in the state plan, obtaining economic powers equal to those of a province, and gaining control over the four dozen or so large industrial enterprises decentralized (*xiafang*ed) to city management in and after autumn 1984.[11] Not surprisingly, this measure has not rested easily with Hubei Province, which felt threatened by each dimension of these changes. Indeed, during the period surrounding the granting of this new status to Wuhan, it was clear from press material that there were tensions here between province and city.[12]

But we need to assess the concrete outcomes of the reform, both financially and behaviorally, in order to understand the extent to which, and the manner in which, this step has enabled Wuhan actually to gain new power. That is, we need to know whether Wuhan does, in fact, now have more autonomy, what sort of autonomy it is, and what are the costs and benefits such autonomy brings at this point. For the purpose of *jihua danlie* was to "make central cities step out of the small world created by administrative demarcations,"[13] and, particularly in the case of Wuhan, to give that city "fuller scope to play its role as a major city."[14]

First of all, it seems that the reforms have not really done all that much to give Wuhan the capital base it needs for large-scale change.[15] Despite a shift in the direction of the flow of funds between Wuhan and Hubei beginning in 1985, the amounts of monies going between these two levels and the central government has really changed very little. Though Wuhan can now retain about 20 percent of its local revenue, which is a little higher than before, this retention rate is the lowest of all the cities that now have provincial status in the state plan.[16]

Moreover, the center now subsidizes Hubei for the approximately 100 million *yuan* it lost as a result of the *xiafang*, while Wuhan now gives the center about 100 million *yuan* more than Hubei used to give to the center, when it was Hubei that benefitted from the profit and tax collection coming from the enterprises now *xiafang*ed. It seems that financially the principal advantage to Wuhan is that, since its fixed ratio for retention is now a little higher than before, and since it is now the collection point for the revenue from the decentralized factories, its incentives have increased. And, correspondingly, Hubei's loss is more in this gray area of potential reward for enhanced activity.

The argument those in the city made, both on the eve of reform[17] and in the midst of its implementation[18] was that, as the city's economy becomes stimulated, it will go on to invigorate the entire provincial economy. This could happen, local proponents held, given Wuhan's superior transport and commercial advantages and the possibilities for these systems to set a chain reaction in motion.

Still, it appears, Hubei feels cheated by the reform. Consequently, the always ambivalent relationship between the province and its premier city—one marked from the start by paternalistic protection but also by pressures for Wuhan—has now become even more complicated. Most blatantly, as recently as early 1987, Hubei withheld resources over which it still has control from Wuhan enterprises, acting on the premise that the province no longer stood to gain from their successes.

The items involved were cotton, the great bulk of which (95 percent according to a spokesman at the Wuhan No. 1 Textile Mill) is still directly allocated by the plan, and electricity, allegedly now under the control of a newly organized Central China Electricity Network. With the creation of this network, Wuhan's electrical power should in theory come from that organ's administrators directly. But, according to a local economist, "Whether we can realize this change depends on the relations between Hubei and the Network."[19] The interesting point here is that Hubei's power is evidently still so much greater than Wuhan's own that if Hubei's *guanxi* (personal relations) with the Network's management is good enough, and if it chooses to apply pressure, it could work to override Wuhan's rights under a new allocation system specifically designed to ensure Wuhan's allotment.

The opening of all sorts of new circulation channels with the reforms mean that Wuhan can now go outside the province for the materials withheld by Hubei, but at a cost: cotton obtained farther away, for instance, means higher transport charges. Steel, petroleum products and home appliances from Wuhan can be bartered, but that puts a drain on these valuable local barter chips that the city could otherwise be using for items the center never allocated to it in the first place.

One final area where Hubei still can hold sway is industrial deployment. It seems that Hubei is resisting placing new firms in Wuhan, and that arguments over this occurred in Beijing in the first half of 1987. Though feasibility studies could presumably settle the issue, personal factors can influence such studies, just as they affect every other aspect of economic life in China.[20]

As for the nature of working relations between the two echelons now, there appears to be some confusion as well. According to the tenets of the *jihua danlie* arrangement, Hubei no longer has an economic, but only an administrative, connection with Wuhan in hierarchical terms. Economically, the two get together simply on a business (*jingying*) basis, but not as superior and subordinate.

That is, since investment is no longer allocated through the province for Wuhan, but comes directly from Beijing, Hubei's formal means of exercising financial control over the city have terminated. But administrative powers still in Hubei's hands include the crucial one of personnel control, and, as one now highly prominent city factory manager admitted, "Economic and administrative affairs are hard to separate."[21]

Indeed, though Hubei now no longer approves the projects Wuhan undertakes, and though the city does not report on its activites to the province anymore, the expectation that Hubei will provide support of various sorts continues to exist. There are still occasions on which Hubei's offices provide suggestions, market information, and even grain, non-staple crops, and raw materials, and offer help in dealing with such matters as the disposition of economic criminals.[22] The fuzziness of the relationship came through in a discussion with a man from the Financial Reseach Institute of the People's Bank's local branch office. At different moments in the space of one interview he agreed that the province supports an eight-city intra-provincial financial network centered on Wuhan, still exercising a lot of "leadership" over it, but then later explained that capital outside the state plan is not under Hubei's control, and that, since such capital is what is used in the short-term capital markets in this network, Hubei has no power over it.[23]

Taking together all of these factors—the continuing capital shortage, the persistent control by Hubei over some really crucial resources, and the combination of support and control that the province still extends over the city—it is fair to conclude that Wuhan is not yet a truly autonomous center of economic activity.

Decentralized Enterprises

Probably the most significant dimension of *jihua danlie* should be the decentralization of enterprises from provincial to city management that began in the fall 1984. About 50 firms fell into this category. The motive behind the *xiafang* was to enable cities to unify the management and organization of the production and circulation activities occurring within their borders, and ultimately to permit urban centers to encourage more specialization among their firms.[24]

The process of decentralizing has not proved easy. One source held that, as of late June 1987, not all of the firms had been decentralized;[25] a city economist revealed that the *xiafang* was not a thorough one in all cases;[26] and cadres at the Wuhan Heavy Machinery Plant (Wuzhong) explained how the process in their plant, begun in January 1985, was not completed until August 1986. The sticking point, apparently, involved settling the division of finances between levels.[27]

Interestingly, the three *xiafang*ed firms that I was able to visit exhibited a different pattern of decentralization in their management and thus a different relationship to the city. One (Wuzhong) chose to remain within its old, tried planned relationship; one (Wuhan Steam Turbine & Generator Plant) is branching out, so its manager says, entirely on its own, to become the head of several "socialist trusts," one nationwide (Yangtze Energy Corporation) and one within the Central China Region; and one (Wuhan Iron and Steel Works, Wugang) is too huge for Wuhan to administer, but is now more available for forging the sort of barter relationships with city firms that once took place on a more ad hoc basis but which have now become formalized.

In the case of Wuzhong, the plant attended a meeting at the Ministry of Machine-Building in Beijing 1985, at which it was given a choice: thenceforth, it could depend upon the

Ministry, the province, or the city to procure for it the raw materials that constituted its base, guaranteed amount under the state plan. Because of the greater certainty of supplies coming from the center, the firm picked the Ministry of Metallurgy, whose office in Wuhan (which has no connection with the city government or its local Machine-Building Bureau) thereafter sent the plant its allocation. In this case, thus, the only real difference from before in its source of supplies was that the plant ended up drawing them from the center through its own choice, and not as the result of a compulsory plan.[28]

The original Wuhan Steam Turbine & Generator Plant had expanded itself into a trust called the Yangtze Energy Corporation by the time of my visit, as well as creating several smaller such groups.[29] This company, established in January 1986, after the *xiafang*, had recruited member firms nationwide on a competitive basis, to a total of 59 by the time I arrived. The corporation had factories responsible for every phase of production, from design to installation.

For this, as well as his other trusts, Manager Yu Zhi'an claimed there had been neither economic help nor interference from either Hubei or Wuhan. However, he did admit to a vague "support" from top city government officials, and in China political facilitation can go a long way. As a Wuda professor put it, "in China political support is economic support."[30] Nonetheless, the upshot is that the city has not really enhanced its management powers over this firm and its activities.

Of the three, it was only in the case of Wugang that the city truly changed its situation with regard to one of these decentralized plants. Of course, everyone queried quickly commented, Wugang was far too massive for the city to handle. As the City Planning Commission's officials explained, the province could not manage it, and so Wuhan cannot either. Only the center is competent to dispose of its sales and guarantee its material supplies.[31]

But an exchange has been arranged between the city and the plant, by which Wuhan agreed to provide services to the firm's employees, such as housing, schools, and shops. In return, the plant ensures the city of one percent of its output, which is sold to the city at state prices. In addition, Wugang has formed a merger with a number of small city steel factories that had been operating at a loss. According to this deal, these small plants process parts for Wugang, and the plants can count on a guaranteed source of steel.[32]

In forming this merger with 21 plants manufacturing such items as rolled steel, steel sheets, strip steel, coke and refractory materials, Wugang took over the direct control and unified management of these firms which previously belonged to a city metallurgy company and a local mining area. In this way, Wugang was at once able to expand its own capacity, help out small factories that lacked raw materials and the funding to modernize themselves, and solve a portion of the city's deficit enterprise problem. In this one instance specialized production did result from the *xiafang*, even though Wugang remained essentially under Ministry management.[33]

Such exchanges as Wugang's occurred in the past in Wuhan. Notably, in recent times, during the readjustment period of 1979-1982, local heavy machine-building plants were desperately short of supplies and bartered such products as valves for Wugang's steel.[34] But unlike the valve deal, which had been more covert, this new one was approved by the city's Party Committee and the urban government.[35] So in this case too what seems to be innovation when it goes under the name of "reform" is actually an adaptation and formalization of a type of relationship that grew up first outside of official channels in a previous period.

Overall, then, *jihua danlie* and the accompanying *xiafang*

CITY, PROVINCE AND REGION

of local enterprises present a complicated picture. Though there is some financial return to the city, it falls far short of providing for the sort of major transport and communications construction the city desperately needs in order really to realize its autonomy.[36] On some occasions it seems that all this has only aggravated Wuhan's relations with Hubei. Perhaps at such times the city finds that the old superior/subordinate bond may in some ways have been a more convenient one than today's more purely business ties. The degree of ambivalence local people feel toward the changes was evident in one remark by a local economist in particular: *"jihua danlie you haoqu...buneng shuo shi meiyou yidian haoqu"* (*jihua danlie* has good points...you can't say that it is without any good points at all).[37]

Certainly the enterprise decentralization remains a long way from the point where Wuhan can truly organize the enterprises within its borders for purposes of its own. Looking just at my material on this one theme, Wuhan's powers still remain circumscribed, if in different ways from the past. Thus, all in all, there is at best only an incipient basis for the city to form the core of a regional economy building on the enterprises supposedly newly under its jurisdiction.

2. Enterprise Autonomy: Bankruptcy, Leasing, Manager Responsibility Systems, and the Bureaus

The much publicized movement to give enterprises power over their own productive processes and responsibility for their profits and losses—as the Chinese put it, to separate ownership power from management power—finds its expression in an extremely limited experiment with bankruptcy, a larger-scale program of leasing small state-owned firms (mostly in the commercial system) to their staff and workers, and in an array of factory and store manager responsibility systems.

Often it turns out that in the model cases where these

reforms are going on, old arrangements are being recast so that management and workers get new incentives, while the essential relationships between firm and bureau have not been grossly altered. Success seems dependent on political facilitation for these firms in particular, where management bureaus seem to have been explicitly forbidden from intervening, even though such intervention is still rife in ordinary, non-trial-point enterprises.[38]

Bankruptcy

The bankruptcy experiment is, at this stage, still nothing more than a demonstration, and apparently is heavily laden with political elements. As of early 1987, only 11 state and collectively-owned firms in four test cities (Wuhan, Chongqing, Shenyang, and Taiyuan) had been targeted nationwide for trying out this system and given yellow warning cards indicating imminent bankruptcy. Of these, only one, in Shenyang, actually went under in August 1986. These firms were warned that unless they reduced losses within one year, they would be declared bankrupt.[39]

But the draft law passed by the National People's Congress Standing Committee in December 1986, shows the heavy role still reserved for the failing firm's management bureau. For it is the bureau which makes the application for bankruptcy; whose agreement the creditor must secure before it can appeal its claims; which has three months to find a way to straighten up management (*zhengdun*) within the firm and two years to carry out this adjustment; and whose personnel are subject to administrative penalty should the enterprise fail.[40]

In Wuhan, the No. 3 Radio Factory was the first to recover of the three ailing factories in the city picked for liquidation if they did not reduce their debts within a year. Press material gave the credit for its revival to a change in management, plus heightened effort by the workers in response to the threat of closing down.[41]

But when an official from the City Planning Commission was pressed, he agreed that any manager requires support, help and cooperation in order to make a turn around, and that can only come from bureau leadership in the form of capital, loans, and raw materials.[42] So at this early stage, enterprises are selected to "go bankrupt," and it seems that whether they in fact do so is still a matter of political choice.[43]

There are a number of other options from which political decision-makers and local firms can draw to deal with the pervasive problem of deficit.[44] One of these other possibilities for a failing firm is to form a merger with a large, successful plant, such as in the case of Wugang's arrangement with the 21 little local factories noted above.

Another smaller-scale version of this was reported in the city paper in June 1987: a state-owned towel factory signed a one-year lease contract with a street-managed thread enterprise which was losing so much money that no wages had been paid for three months prior to the takeover. Management powers went to the towel factory, while the thread firm paid monthly rental fees.[45]

This public, formalized union is apparently being copied spontaneously in other firms. Just by chance, I witnessed this in the midst of one of my interviews at a national model factory. A staff member interrupted us to inform the manager that a nearby firm under the same industrial bureau, about to go bankrupt, had come to appeal to be bought out.[46] These other solutions indicate the flexibility and innovation now present in the system, but also serve to underline the disinclination to allow bankruptcy to become a reality.

Leasing and Manager Responsibility Systems

Leasing, which requires official sanction, is yet another possibility.[47] By mid-1987, over 10,000 small-scale industrial and commercial enterprises had been leased nationwide[48]; in Wuhan, 120 small-scale state-owned industrial firms were leased by summer 1987, constituting half of the number of such firms in the city, while more than 300 (about 44 percent) of the commercial enterprises in the city were leased, including both state- and collectively-owned ones.[49]

This reform entails the lessees—usually the manager and a group of top staff personnel—putting down personal property or cash as collateral and losing it or getting it back, plus a bonus, depending on how well the contract (which stipulates a profit target) is fulfilled. Alternatively, staff and workers' wages can be supplemented by bonuses or diminished in accord with the amount by which the firm fails to meet its target.[50]

Though the staff and workers who lease their firms do obtain the right to operate state property according to decisions they freely make, they are enjoined from dismissing the original workers "at will." Also, these firms in the end remain under the charge of their old administrative departments.[51] A report from a forum on invigorating small state enterprises in Wuhan, held in late April 1987, is very revealing. It notes that, in some instances, "administrative interference still exists...we need to 'open the bird cage and release the sparrow.'"[52]

In distinction from these solutions for failing firms, the new manager responsibility systems[53] tend to be placed in larger enterprises and stores, and where the management has been successful.[54] Where management may be switched in bankrupt-prone plants, this probably will not happen in firms using the manager responsibility system. Even where elaborate bidding and examination schemes were installed to select new leadership, the previous assistant party secretary of a large department store can emerge as the new manager from a competition of over 40 applicants.[55] This system is also distinguished from leasing in that enterprise income, not mortgaged personal property (as in leasing) is forfeited where there is a loss.[56]

In most cases, the contract—whether for leasing or for manager responsibility—is meant to replace the old state plan for the firm.[57] Though it holds for three to five years, not one as the plan did, its specifications are some of the same targets as those that were once the contents of the firm's plan. These include profit, gross value of output, quality, and growth of fixed assets, or sales volume in the case of stores. These contracts are signed by the firm and its management bureau, which acts as the representative of the owner (the state), after the bureau has formulated the contract with the city's economic commission, the bank, and the tax and finance bureaus.[58]

True, there are fewer overall stipulations than the plan had, and there is more leeway on some matters—such as in new product development, finding supply sources and choosing sales outlets. But one informant made the interesting point that his plant's four-year contract actually firmed up and stabilized its old production targets, which used to shift a lot even within the space of a year, and thus used to lead to confusion.[59] Another admitted that the city's Finance Bureau supervises the contract's implementation on a monthly basis to make sure the factory is operating within the scope of state regulations.[60]

Moreover, another factory's personnel hinted at disagreements with its bureau over the selection of new product development, even though this power now belongs to the firm, decisions in this sphere not having been written into the contract. In this case the interviewees maintained that in such disputes the bureau usually lets the firm do what it wants.[61] Often enough, though, according to the press as of summer 1987, departments in charge of enterprises were "still focusing on requiring enterprises to fulfill the main economic targets," despite the institution of contracts.[62]

If they do not interfere as much as before in these now contracted-out firms, the bureaus can still ensure that their firms do well, thereby facilitating the success of these experiments. For, despite the manager's new responsibilities under these contracts, it seems that the firms go on relying rather heavily on market information supplied them by the bureaus, which have access to far wider networks than does the typical enterprise itself. Depending upon the principal raw materials that the firm requires, the bureau can continue to be a crucial supplier.[63]

But where no special experiment is underway, the bureau has definite incentives to continue to intervene. For, in the first place, the statistical system has not kept up with the management reforms, so that the bureau is still held responsible for the enterprises' financial achievements. Each year it must report to the city government on the taxes and profits all of its firms have earned. In addition, the bureaus remain dependent for some of their expenses on the management fees (*guanli fei*) the firms must pay them yet. This important source of their funding reinforces the tie of interest between the bureau and the enterprise's success. Until these bonds are broken, the new reforms can only hope to bear fruit at selected, showcase sites.[64]

In sum, the reforms in enterprise autonomy, while certainly a step toward cutting the controls the administrative system has historically clamped on the firms, cannot succeed on a large scale at the present. This is not to say that there has been no reform, nor that there cannot be more of it. Its strengths so far lie in using but loosening old arrangements—in this case, the state plan—for new ends as contracts with fewer, but some of the same, stipulations now guide enterprise behavior.

In addition, those who produce the plant's economic results in the trial firms—if their bureaus let it happen—will have to suffer financially if their firm does, but may benefit as it thrives. This "if," however, is central to the scheme at this phase, in which political support is being used to demonstrate possibilities. Where there are those possibilities, as we will see below, exchanges are taking place that supercede the old administrative arrangement and the territorial boundaries within which they were confined.

3. New Investment Sources

The creation of new sources of investment in the last few years, to supplement if not to supplant the pre-reform planned pooling of funds through banks and financial departments, is crucial for invigorating enterprises. In Wuhan, the problem of capital was considered an outstanding issue in economic construction in late 1985. Causes lay in a slowdown in the circulation of capital because of the readjustment of the early 1980s which suspended, merged or converted many firms; in backward technology and outdated equipment; and in a batch of new enterprises desperately awaiting funds before they could develop their productive potential.

It was estimated that, as of 1985, some 300 million *yuan* was needed in the city for the renovation of fixed assets, while another 230 million was required for technical transformation. But, according to one source, only a portion of all this was then available through the usual channels. Meanwhile, rising income in enterprises, other units, and the working population in both the city and countryside was leading to an accumulation of idle funds that had no outlet other than the purchase of consumer goods.[65]

To deal with this problem, a nationwide one, new forms of fundraising have been endorsed and are under experimentation. These include raising funds by issuing shares (usually by selling stock in a firm to its own staff and workers), bonds issued by banks, and, most recently, opening stock markets in a few trial point cities. Other methods are through joint operations, as when enterprises invest in each other, or when they exchange scarce raw materials or products for capital from the other party.[66]

Yet one more channel is the short-term inter-bank lending, begun in 1986, that constitutes the first step in the formation of capital markets. That program was launched in five cities, including Wuhan, and then was extended to others. By early 1987, inter-bank loans amounted to 30 billion *yuan* nationally.[67]

Here is a type of reform that really has no precedent in the old system. But it is one which, though proceeding haltingly, is still stymied to some extent by the *tiaotiao* (roughly, vertical lines of authority) of the bureaucratically organized banking of the old administrative system. Also, its scope of operations at this point is still largely quite localized. To play their intended role, capital markets must eventually really cut across city and provincial borders, and be open to economic and not just administrative regulation.

Shares, Stocks, and Bonds

In Wuhan, a city Financial Trust Company (*jinrong xintuo gongsi*), set up in 1981, constituted the spearhead for reforms in the financial sector. Its job from its initiation was to raise funds for local economic construction. This is done by acting as a broker for companies and factories in the city. Its list of subscribers include academic and research institutions, factories, organizations and even individuals.[68] But mainly it brings together capital-short firms with others that have idle funds they can use to subscribe shares in the former. It also helps to direct such funds into urban construction projects, public facilities, and public transport.

By the end of 1984, the company had raised 101.6 million *yuan* mainly by issuing shares, and had granted 66.98 million in loans. In 1985, savings in the city increased by more than 400 million *yuan* over the year before. In that year the city organized enterprises to issue stocks and bonds to the general population, staff, and workers in the enterprises, collecting over 100 million *yuan* by doing so.[69] As of mid-1987, only four enterprises were directly issuing their own shares, while nearly 1,000 firms were issuing bonds. At that point, "society" held 400 million *yuan* worth of bonds and shares, with the overwhelming majority of bondholders being firms and individuals resident in the city.[70]

As early as 1985, problems had already surfaced in Wuhan's money market. Illustrative of these were cases of firms using dividends as disguised bonuses, thus evading taxes on the bonuses; seeking funding without first doing feasibility studies, with the result that very few people bought the shares put up for sale; collecting funds that never got used; only soliciting funds for investment, but neglecting to consider production costs; and setting the dividends too high. By autumn of that year, Wuhan had already established a management group to control the issuance of these shares.[71]

In spring 1987, in the midst of a period of nationwide retrenchment of economic reform, the State Council published regulations to control the sale of bonds by enterprises.[72] The purpose was to try to ensure that the collection of funds would serve state goals and guarantee the construction of key state projects.

The ruling placed the People's Bank in charge of issuing all enterprise bonds. Thenceforth any effort to raise funds for investment in fixed assets would require approval by departments concerned and incorporation into the funds under state control explicitly set aside for this purpose. Moreover, state-owned firms were enjoined against selling shares, and collective firms could only issue shares among themselves but could not sell shares to the public. Criticisms of extortion and of blind construction accompanied this call for strict supervision.[73]

Wuhan took advantage of the continued permission to

offer bonds even if under strict bank control in late May 1987. At that time the Hankow Branch of the People's Bank put up for sale bonds with a time limit of one year at a nine percent interest rate.[74] Then, at the end of June, Wuhan became the third city in the country to open what it is calling a "stock market," after Shenyang in August 1986, and Shanghai in September 1986.[75]

The Wuhan Trust Investment Company under the Industrial and Commercial Bank was given the task of managing this market.[76] By then the city had issued various kinds of long-term bonds, stocks, and a large amount of certificates of deposit, including various large and middle-sized enterprises' stocks and bonds, altogether reaching a value of more than 800 million *yuan*.[77]

But the continuing controls, both administrative and economic, that the People's Bank still exercises over its specialized component[78] limits the new stock market's ability to function as a genuine site of exchange.[79] And, oddly for a stock market, local observers admit it does not promise soon to be a place where the actual transferral of stocks will occur. Moreover, the paucity of capital in the city seems likely for the present to confine the activity there to local buyers and sellers.[80] In the words of a local economist, *"Ni meiyou shili, dui waidi buxing"* (If you don't have any real strength, it won't do to go outside [the city]).[81]

Two other forms of funding are coming into wider use in recent years, and these also aim to break out of the restrictions imposed by the old system of administrative hierarchy. One of these is illustrated by the Wuhan Bazaar Share Group's investment in the Yangtze Aluminum Products Factory's pressure cookers. According to the agreement, the factory will give the store 100,000 cookers over a three-year period at a preferential price, regardless of demand. In exchange, the store has invested two million *yuan* in the factory as a kind of interest-free loan.[82]

As in other instances, political facilitation sparked this reform. The city made this store (the Wuhan Bazaar) a keypoint for reform, when it extended permission to draw up the share group, the only one of its type in Wuhan. Set up in late 1986, the group, which offers a 15 percent dividend on shares (i.e., twice the rate of bank interest) had already collected 10 million *yuan* in shares by mid-1987, all drawn from partners picked by the store.

Its seven investors include Wugang, an insurance company, and the local branches of the People's Bank and the Agricultural Bank. Since the risk was smaller with local partners whose business acumen was known to the store, and in the interest of developing the local economy and thereby boosting purchasing power there, the store intentionally invited only units from Wuhan to join.

And the last type of non-administrative funding is foreign exchange, the disbursal of which, however, in most cases still is tightly controlled by the People's Bank and secondarily by the management bureaus of the firms. In the tiny handful of privileged experimental reform firms, though, such as the Wuhan No. 2 Bicycle Factory, managers get full authority over their own imports of foreign technology in return for having successfully exported.[83]

Offering stocks, shares and bonds is a valuable means for reorienting the structure of funding in Chinese enterprises. To some extent these practices are already underway; but many elements of the old system continue to keep the scale of such business small and the scope localized. These elements include capital deficiencies at the urban level; the state's concern to supervise the use to which funds are put; ongoing management by the People's Bank and its branches; and the tendency of even experimental units, given freedom to pick partners, to stay within the confines of the local economy. Here then we see that even political facilitation does not yet necessarily override the persistent presence of features and organs from the previous era.

Capital Markets and Networks

Wuhan historically had played the role of financial center in the Middle Yangtze Region. At one time Hankow boasted some 230 traditional and modern banks, and was one of the nation's three large capital markets. Its interest rate, foreign exchange rate, and gold and silver prices influenced the neighboring districts.[84] Its geographic position, convenient transportation and developed communications, scale of commercial exchange and economic strength in industrial growth all make it a favorable site for the development of a financial center again in this period of reform.[85]

Indeed, in commercial credit, regional banking, and short-term inter-bank loans the results are promising, according to a local economist.[86] By the end of 1986, nearly eight billion *yuan* had been traded in four different capital market networks and four special capital markets, all centered on Wuhan.[87]

In April 1987, a fifth short-term capital network based in Wuhan formed.[88] This one comprises 55 cities that form a "financial cross" in the Yangtze Valley and along the Guangzhou-Beijing Railway, and the Construction Bank acts as its agent. Its particular mission is to collect loans to support national keypoint construction.

The first of these five networks to be established was one of seven cities within Hubei province; the second involves 11 cities (the first seven *jihua danlie* cities, plus Shenzhen, Zhuhai, Changzhou, and Haikou); the third is of 27 cities along the Yangtze, including Shanghai, Nanjing, and Chongqing; and the fourth has 33 Central-South cities that are not along the Yangtze, and is managed by the Industrial and Commercial Bank.[89] These networks offer loans usually about a half year in duration, to be used for circulating capital rather than for fixed asset purchases. They also provide for the exchange of economic information.

Although the cities in the 11-member network have the most economic strength and are subject to the least administrative interference, in fact it is the web within Hubei that sees the most action. For their communications are easiest and exchange is frequent, if not in large amounts. But, as noted above, the research official from the People's Bank who discussed these networks was ambivalent about the extent of Hubei's control over this market.[90] This suggests that some administrative interference from the Hubei branch of the People's Bank is probably present, which would limit the market's ability to respond freely to economic forces.

This inference derives support from the comments of a local economist that Wuhan cannot really be counted yet as a financial center in the true sense of the term. For, he pointed out, the banks located and doing business there are not yet enterprises, and so, as the bank's researcher admitted, they lack the necessary independence for economically based decision-making. In the same light, the participants in the various capital networks are branches of the People's Bank and of the specialized banks linked to it, all of which must answer to their administrative superiors in the banking bureaucracy.[91]

4. New Circulation Methods

Lateral Exchange

Lateral exchange (*hengxiang lianxi*) is the name for the new, supposedly economically based transactions developing in China since the reforms began. Its purpose, just as with all the reforms, is to connect supply sources with demand, and to demolish the blockades between districts imposed by the plan and its administratively structured marketing. In the urban environment, in particular, the aim has been to create economic liaisons between enterprises and between regions centered on cities as the financial and economic management powers of cities grew.[92]

These new partnerships serve a myriad of disparate functions, depending on the specific needs of the firms that form them.[93] Some enterprises join them to gain a guaranteed source of supplies and parts. Thus, Wuzhong obtains a fixed supply of iron from a Hunan factory; the Wuhan No. 3 Printing and Dyeing Factory gets some of its cotton from such cooperative arrangements. The Wuhan General Machinery Plant and the Wuhan No. 2 Bicycle Factory both placed themselves at the center of "enterprise groups" (*qiye jituan*)[94] that ensure each of them needed parts: the former gained expanded capacity by pulling together failing firms under its own aegis; the latter, after bidding, formed a liaison with the nation's premier bike firms in Shanghai which send it components.

Others draw on these deals for locating work or making sales. The Wuhan Chemical Machinery Factory, which is mainly in the business of processing steel, uses these unions for finding jobs. Another similar case is Wuzhong, which is linked to a company that sends it equipment to repair. And the Wuhan No. 2 Bicycle Factory has an arrangement with commercial bureaus in several major cities that, in essence, give it three-year interest-free loans in exchange for a shipment of bicycles. Both sides profit, since the plan's old intervening wholesaling companies are cut out of the picture.

The Yangtze Energy Corporation has formed several enterprise groups, as noted above. These "socialist trusts," as their manager calls them, pool investment funds from their members which are donated to the central firm in the group in exchange, for equipment, electrical power, and for activities—such as building a reservoir. A union made up of the wool-producing firms of Wuhan constituted itself as a company. These textile plants hope to enhance specialization and improve their organizational structure and to allocate more rationally the capital and raw materials within the local trade.⁹⁵

In the course of implementing this reform over the past several years, to what extent and in what ways has the initial intention taken shape? That is, has free exchange arranged according to comparative advantage, really begun to replace the plan? Or are prohibitive obstructions from administrative organs stopping it up? Or is there yet another way to understand what is occurring?

Available statistics do not provide a clear answer. In early 1985, *Xinhua* reported that 20,000 interprovincial economic and technical cooperation contracts had been signed in 1984, an amount that was two times the figure in 1983.⁹⁶ By the end of 1986, over 32,000 integrated organizations with lateral economic ties had formed into 24 lateral economic liaison networks.⁹⁷ But this gives us no idea as to the nature of these unions.

Nor does it indicate the degree to which combinations form within provinces and more local areas—thus, continuing to respect the old administrative boundaries—as opposed to those of a cross-provincial nature. Wuhan, for example, reported in 1985, that enterprises there had built up 517 economic associations and cooperative organs with its suburban counties and had signed over 300 cooperative projects with more than 20 counties and cities of Hubei—but without telling how many ties it had established outside the area.⁹⁸

Yet another source states that in 1986, 1,140 enterprises in the city engaged in such cooperation (termed *lianhe xiezuo*, or united cooperation) with a total of 2,593 firms and units in 28 provinces and municipalities, signing 3,324 cooperative projects involving 251 million *yuan*, but this time without saying the number of local projects in that year.⁹⁹ While these figures give the impression that the bulk of these exchanges are national in scope, qualitative data from interviews with Wuhan firms, governmental departments, and economists would suggest the opposite.

But is this because of resistance from administrative units? In many cases, it is. A Wuhan University economics professor explained that in the last few years these exchanges have developed quickly, and pointed out that both academic economists and governmental bureaucracies agree on the policy. But, he also noted, the State Planning Commission wants to put some 500 of the largest enterprise groups under its own control, and a lot of the unions of all sizes are suffering from administrative interference.[100] The press seconds his views, as in this example: "If state planning departments can grasp about 100 very large industrial groups, that will help in promoting the organic combination of plan and market...."[101]

The press, from the outset, has been bemoaning the "conflicts" that "always arise between planning and cooperation." One analysis among many that are similar explains how cooperative undertakings, not provided for in the state plans, are eschewed by many units, since the fruits of their investment will not accrue to their mandatory target fulfillment.[102]

There are also many stories of horizontal (as opposed to vertical, administratively arranged) contracts being forbidden by management departments (bureaus) which resist the outflow of funds and raw materials under their control and the transfer of key technology that might cause them to lose their competitive edge.[103] In Wuhan by 1984, 35 enterprises were already reported to be getting their spare parts through bidding. But when the scheme was just beginning in 1982, some former parts suppliers of the washing machine company involved tried to prevent the switch.[104]

There are also cases where political facilitation and the plan's bonds themselves are permitting the exchanges to go forward. In such cases the bureaucracy is no impediment, but is, in fact, promoter and expediter, if not always intentionally. Where the promotion is purposive, it may take the form of assistance from top local party and government officials who are creating showcase models.

Thus, the manager of the Yangtze Energy Corporation, for instance, claimed to have no interference in his activities, and no economic support from the state, as mentioned earlier. But, being the party secretary of a major, once centrally managed state enterprise, his personal connections with the mayor and the city's Party Secretary are strong enough that the "interest" he admitted that they have showed in his firm must have translated into having his way smoothed in the face of any opposition.

Similarly, the Wuhan No. 2 Bicycle Factory, a national model in bidding for parts, was selected for this project by the City Economic Commission; the decision to carry it out was announced (i.e., at least endorsed, probably backed) by the still-powerful ex-Mayor Li Zhi.

On a more explicit level, the city has set up two crucial committees to further exchanges of which it approves. One is an Economic Levers Adjustment Committee (*jingji ganggan tiaojiehui*), composed of representatives from the city's tax and finance bureaus, the People's Bank, and the Price Commission. This unit is empowered to give favorable treatment in pricing, taxes, and bank loans. For example, it ensures that there will be no income tax on the parts supplied by the minor partners of the enterprise groups, which become workshops of the larger firm. That larger firm alone pays a value-added tax.[105]

The second, equally significant, body is the Wuhan Economic and Technology Cooperation Committee, set up in June 1985.[106] This organ is designed specially to promote lateral exchange and to help solve any problems that may arise in the process. When feasible, it encourages firms in the same trade to combine, brings them together to negotiate over terms, and supplies them with relevant market information once their union is achieved.[107]

But the existence of these committees has a double edge, sometimes supportive but also at times restrictive. They clearly entail a state-sponsored effort to keep a grip on the growing trade. For instance, the primary job of the second organ is to make out a program (*guihua*) for the projects city enterprises join and for the surrounding region. Moreover, it reports to Hubei Province, where a similar committee presides over exchanges at that level, and Hubei maintains a relationship of business leadership (*yewu lingdao guanxi*) over the city in this regard.[108]

Especially interesting is the background of the 30 odd personnel who staff this committee. All of them previously worked in the area of materials exchange, and so are quite familiar with the situations and products of the enterprises in the city. Some were employees of the Materials Supply Bureau in the past; others were agents delegated by the city government (planning or economic commissions, or the city government's office) to get market information to supplement planned supplies. Yan Qingfu, now the head of the Committee, had been Deputy Secretary-General of the city government before the formation of this body.[109] Here then people and bureaucracies held over from the past are in a position to facilitate at least as much as to obstruct reform.

The Materials Supply Bureau also aids firms in need of raw materials whose planned allocation was cut drastically with the advent of "guidance planning." One beneficiary is the Wuhan Chemical Machinery Factory, which now receives less than 10 percent of the steel it needs from the state plan.[110]

Similarly, Wuzhong's extra-plan activity depends on a company directly derived from the former Second Ministry of Machine-Building, now entitled the Chinese Machine Tools General Company. This body was established in 1982 as an economic, not an administrative, entity. It manages the sales of its member firms, all of which are factories originally subordinate administratively to the Ministry.

Wuzhong also gets repair jobs as a result of its membership in the Wuhan Machine Tools Repair Associated Company, which is a subunit under the larger company. Its leading staff were all employed by the Second Ministry of Machine-Building, so they know what equipment every factory, nationwide, has and what it might need in the way of repairs.[111]

Turning these administrative agencies into business firms helps to coopt them into the reform process. It is clear that they have been given crucial new responsibilities that virtually set much of the "market" activity in motion. In fact, the Materials Supply Bureau's personnel now receive an economic payoff, in the form of commissions, for doing so.[112] Some have been able to go even further, turning this new business mode to illicit advantage, rerouting materials that could be exchanged without their assistance and extorting "management fees" from the process.[113]

Information flows, still heavily dominated by state-run organs, are also expedited by those in the know from the past.[114] Here, again, the local Materials Supply Bureau plays a role, for its cadres have relationships with factory managers across the country, many of whom have been on the job for years.

But the most significant actors are the old factory purchasing agents (*caigouyuan*) who served their firms where the plan proved inadequate. All industrial enterprises have some half dozen to a dozen of such personnel. Each agent specializes in one or more materials about which they have built up a stock of knowledge over the years as to where the material can be obtained. Most of these people are still in place.

They go to work by linking into the pervasive enterprise information networks (*qiye xinxi wangluo*) that stretch across China, of which there are two main types. The major network or *da wang* tells what can be found in particular cities; the little one, the *xiao wang* is more specific, pertaining to the materials held by individual factories. By continuing to communicate through these networks, factories find much of what they ostensibly buy "on the market."

For department stores, the story is not too different. Much of their stock comes from factories whose staff they met over the years at semi-annual national supply meets. These meets, each of which presents a particular type of commodity, were formerly run by the relevant wholesale station. Under the regime of the plan, factories, all of whose output was automatically purchased and allocated to the retailers, participated just to learn about the stores' demands; today they come to do business. The *guanxi* built up over years of attending the same meets stands the stores in good stead today. Spokesmen from the Central-South Large Commercial Building, a sizable department store that opened in Wuchang in early 1985 (almost all of whose staff worked in one of Wuhan's two other big stores previously), estimated that as much as 40 percent of its supplying factories were known to their co-workers from the past. Though the remaining 60 percent of purchasing done with strangers may seem a lot, one might have expected the percentage of old suppliers to be rather smaller, given the institution of several new forms of forging stocking relationships, such as through ads, or through becoming a sales outlet for the new industrial enterprise groups.[115]

In the lateral exchanges too, previous connections smooth the transition. Of the 18 units in the Wuhan Cement Machinery Complete-Set Company (*jixie chengtao gongsi*) organized around the Wuhan General Machinery Plant, 11 are from Hubei Province and five more are from Wuhan. The other two, a design institute in Anhui and a factory in Jiangsu, were both units with which this machinery factory had business ties (*yewu guanxi*) from meetings, the "exchange of experience," and mutual visits under the plan.[116]

The Wuhan Bazaar's enterprise group invested 100,000 *yuan* in a bicycle plant in Changzhou, one with which they had *laiwang* (dealings) in the pre-reform era, so they know their products are high quality.[117] And Wuzhong signed a contract with a Hunanese factory by which it invested three million *yuan* in exchange for a three-year supply of iron at the state-set price. Their supply relationship had a foundation in the days of the plan, where they had frequent comings and goings (*lailaiwangwang de guanxi*). Those shared experiences before taught the plant that the Hunan firm's iron's quality met their specific demand.[118]

Perhaps the key point about all of these relationships is that the heavy hand of the plan still lingers over the world of exchange in China. As a Wuhan economist explained the situation, "in the past we were sealed up (*fengbi*), so we managed work according to administrative relations. As a result, relations with Hubei and Wuhan are the main part of our present exchanges."

When asked what proportion of Wuhan's extra-plan exchange now is localized within Hubei Province, he replied that it was hard to say, given the large number of paths now open for establishing ties. But he made the only estimate he felt he could in saying that probably 30 percent of the goods produced in Wuhan were sold there, another 40 percent were sold in the province, and the remaining 30 percent traveled to destinations nationwide. Of those, he speculated that the majority of such outside destinations were still in the middle Yangtze region.[119]

This material, added together, suggests the following: because of the adaptability of old organs, networks, and personnel, the distinction between plan and market is far less sharp than theorists might presume. For the same reason, although reform is often enough attended by bureaucratic recalcitrance, where bureaucracies can develop a new vested interest by capitalizing on old skills, they need not resist reform in every instance. But in this transitional era, when such adjustments ease a fair portion of the conversion of systems, the bulk of exchange will probably only gradually seep out of its wonted seines.

Materials Markets

As early as 1979, the state permitted firms to sell on the open market production machinery and raw materials left over after they had met their delivery quotas to the state. Soon a dual-price structure emerged—a state-set price for materials allocated by the plan and a sometimes far higher one for those exchanged privately "on the market"—accompanied by inflation and speculation. In July 1984, a State Council directive authorized special urban materials markets in part to try to curb these dysfunctions.[120]

Their purpose is to centralize and regulate the trade in scarce supplies, stabilize prices, promote the output of short-supply items, and urge investment goods out of the warehouses where firms had accumulated stockpiles purchased in a panic. At these markets the parties to the exchanges can negotiate over prices, but at least the presence of the markets brings some regularity to pricing. The centers offer business information, consulting, advertising, communications, food and accommodations.[121]

But this trade occurs within what is still really a sellers' market. As one source put it,

> now when motor vehicles and steel products are unusually tense, if the materials departments sponsor some trade centers, it's only going through the form of a "center," concentrating all that trade outside the plan into one place, to prevent speculation and stop up improprieties. Strictly speaking, this isn't what's

originally meant by a trade center, [so] calling it a sales center is probably more appropriate.[122]

During the Sixth Five-Year-Plan period, one source states that materials worth 223.9 billion *yuan* were bought on these markets. The traded materials included 99 million tons of steel, 837 million tons of coal, 100 million cubic meters of timber and 137 million tons of cement. Collected statistics, according to this source, indicated that these represented one half the country's total consumption volume.[123] A study published in *Economic Management* made a more conservative estimate, concluding that the amount of the means of production traded outside the plan in 1985 represented "not less than 30 percent" of the total amount of the means of production nationwide.[124]

In Wuhan, in July 1984, trade centers and warehouses opened for such things as metals, construction materials, motor vehicles, timber, electrical products, and chemical and light industrial products, comprising 29 categories of over 2,000 kinds of means of production.[125] Nationwide, by early 1985, there were said to be 130 large trade centers across China buying and selling materials above the state quota.[126]

A study of the Wuhan materials markets judged that the amounts of various categories of goods supplied by the plan and the amounts traded on the markets in 1986 each represented about one half of the total.[127] More specifically, the City Planning Commission calculated in mid-1987 that the city requires about 400,000 to 500,000 tons of steel a year, and obtains about 130,000 tons from the plan's allocations. Enterprises then get the rest on their own. About 50 to 60 percent of the steel used in the city is purchased outside the plan, commissioners believe, and at least half of that goes through the official steel market set up in January 1987.[128]

Despite the liberalization implied by permitting surplus materials to be traded as commodities for the first time in three decades, by early 1987, the state acted to impose a tighter measure of control over this commerce. At this time, the government opened seven state-controlled markets for steel circulating outside the plan in major industrial centers—Shanghai, Tianjin, Shenyang, Wuhan, Chongqing, Xi'an, and Shijiazhuang.[129]

An interesting footnote here is a small notice in the August 11, 1987, *People's Daily* announcing that Hubei opened a steel market of its own a few days before "to develop fully Hubei's steel superiority in central China." This would seem to be another instance of the province's continuing jealous oversight of the city.

These new markets came equipped with special incentives designed to draw all the steel that is outside the plan into their arena: enterprises using steel would be exempt from taxes on added value if they sold their products at these marts, and firms exchanging surplus products there for raw materials for use in state-approved projects could get their materials at government-set or negotiated prices.

At the same time, in Wuhan, the city began to run monthly "materials adjustment markets" (*wuze tiaoji shichang*) where all sorts of raw materials could be exchanged between enterprises at prices close to state-set ones. The city government adds materials from its stocks to keep prices stable there.[130] The city also created a market for idle equipment and parts, where enterprises could mutually adjust their excesses and deficiencies.[131] In the case of all these various markets the very concept of trading means of production is innovation and reform, but it is qualified by state attempts to keep it within bounds.

Another theme these markets share with other elements of urban reform is the use of old bureaucracies for purposes of management. Wuhan's 1987 steel market is, in effect, under the supervision of the city's Materials Supply Bureau, increasing further the responsibilities of that bureau in the new reformed economic system.[132] The Wuhan Metals

Materials Company, under the bureau, is in direct charge of this trade. This company had existed in the past to handle supplies inside the plan, but this new role means it has now taken on an added function: it manages the materials outside the plan as well. The main difference from before, however, seems to be that the company only supplied Wuhan in the past, whereas it now draws steel from and offers steel to outside places too.

Trade Centers

More broadly, aside from the special markets set up to handle the means of production, wholesale centers for all sorts of commodities are also meant to be changing the pattern of Chinese commerce. As one analysis of their role described them:

> An important economic means of smashing blockades is to establish circulation networks with extensive connections with cities as centers...forming extensive sales networks in accordance with economic rationalization.[133]

At first, Wuhan proudly publicized its promotion of 90 different such centers.[134] But after expanding to a total of 102 centers in 1985, they were consolidated down to about 30—according to product.[135] The centers were, presumably, planned to rectify the ills of a rigidly bureaucratic commercial apparatus, which, in Wuhan in late 1984, was composed of some 68 different specialized companies.

These companies were said to lead to an artificial severing of commodity production's intrinsic horizontal linkages, and to create irrational flows and blockades, as each constituted its own separate system, limiting the city's function as an economic center. For the companies split the economy into departmental and local ownership systems.[136]

In contradistinction to wholesalers within that old system, these centers within the new one were to be open to everyone, private or public, and to goods from all parts of the country, using many forms of sales operations, allowing the producers and sellers to meet, and working with flexible prices. In short, it was to be a new circulation system, open in form, many-channelled, and with few links. The new centers, besides, are enterprises, not administrative units, and work according to market-made prices, not state commands.[137]

In fact, even though the centers do handle trade outside the plan, and are much more open as to participants and wares than was the old system, here again is continuity.[138] For most of the centers are run by the old wholesale companies, with the self-same personnel. To quote a local analyst: "[With one exception] all the other trade centers are old wholesale departments (*pifabu*) with changed nameplates (*paidz*)."[139] Moreover, the centers are managed at the city level by the Commercial Administrative Commission, an organ formed in 1984 as a semi-independent company from the former First and Second Bureaus of Commerce and the city's Food Grains Bureau.

But their goods come from factories from many places, with only half or less (40 to 50 percent) of the varieties in them produced in Wuhan itself.[140] In the same vein, already in mid-1985, the *Beijing Review* reported that only 25.5 percent of the centers' clients came from Wuhan, another 21.2 percent were residents of other parts of Hubei, and as many as 53.4 percent were from various parts of China.[141]

While this report gives no clue as to where exactly the marketers hail from, a piece in the local press, in April 1987, claimed that one third of the firms on Wuhan's famous small commodities market street, Hanzheng Street, come from outside, with most of them travelling from East China and the middle Yangtze.[142] Though this does not indicate that Wuhan has become the commercial center for a central China region, it does suggest that its markets draw traders from a wide surrounding area.

All of these reforms aimed at reorienting the bureaucratic

organization of business so that its stuff flows laterally instead of vertically, activated by prices rather than by power, share some common features. Most outstanding is the general principle behind much of this marketing: it often works where it does, because "the plan" and its patterns have prepared the way. That is, much of the marketing of today takes place through the adaptation of bureaucratic organs, the use of tried channels and ties, and the reliance on personnel with inside knowledge of the whereabouts of goods.

Added to this unintentional "help" from the old system, the force of the state is intentionally brought into play at times as well—first to facilitate the success of models, but also to control what are judged adverse consequences when marketers' success outsteps what politicians had in mind.

CONCLUSION

This study of Wuhan's comprehensive urban economic reform has covered a large array of topics: the *jihua danlie* measure of 1984, the decentralization of enterprises, leasing, bankruptcy, manager responsibility, stocks, bonds, capital markets and networks, lateral exchange, materials markets and trade centers.

If there is one dominant theme, it is this: the state plan is still present, more as a grid than as a governor. It is present not just in the sense usually pointed to—that the purveyors of its orders fight to retain their powers. Its existence is manifest in its old offices (even if they now hang out a new nameplate), in the habits it shaped, the patterns of association it fostered, and the *guanxi* developed between those who lived by those habits and patterns. So the plan—through that grid of relationships—works both for and against reform. Indeed, in this human sense, there is no clear or necessary opposition or dividing line between plan and market.

In the case of the movement for urban independence and in the several experiments in enterprise autonomy, the form for change is there—in *jihua danlie*, in the *xiafang* of firms, in leasing—but Hubei and the bureaus still have the means to make an impact on what sort of content fills up those forms. In new forms of investment and circulation, things work best where the old state system intervenes, if in a positive sense, through political facilitation and the use of old organs to serve new functions. But, here too, there has been some obstruction, as in regulations limiting stocks and bonds in April 1987, in banks' continuing vertical control over capital circulation, and in the new state-run steel markets.

The trick, one already being turned, is to shift the economic interest of all the units involved. So Hubei Province has been subsidized for its loss of the decentralized factories, and the bureaus will soon receive compensation when their firms stop handing them management fees.[143] Meanwhile, Wuhan city, and the management and workers in the experimental firms, are all gaining a stake in the new arrangements as the state offers them the incentive to do better on their own.

In all this, though, there is a paradox; for drawing on old information networks, organs and personnel cannot help but influence the formation of enterprise groups, lateral exchange and associations, and extra-plan materials procurement; cannot but shape the business in the materials and capital markets and at the trade centers. Thus, many reform experiments will work better because plan ties, still potent, can facilitate these movements.

But those same ties can also prevent the emergence of totally new formations on a large scale and in all instances at this stage—fully autonomous cities and enterprises (working with neither interference nor support), genuine capital and stock markets. All of these bodies and processes are still somewhat enmeshed in old power systems. Yet it is just such units and markets that must be the building blocks of a new economic system.

NOTES

1. *Renmin ribao* [*People's Daily*] (hereafter, *RMRB*), July 31, 1987, p. 5, translated in *Foreign Broadcast Information Service* (hereafter, *FBIS*), August 11, 1987, p. K12.
2. Here, as throughout, I am dealing with endorsed, legitimated forms of exchange. There is, of course, also a vast and expanding world of corruption in China in recent years that takes off from these opportunities for marketeering.
3. Within a year, five other cities got this status (Chongqing was so designated in early 1983): Guangzhou, Xi'an, Shenyang and Harbin, all of which are also provincial capitals, and Dalian, which is not a capital. Chongqing also is not a capital. *RMRB*, July 18, 1987, p. 1 lists two new cities in this category, neither of which are provincial capitals: Ningbo and Qingdao.
4. *Chiangjiang ribao* [*Yangtze Daily*] (hereafter *CJRB*), Wuhan, August 10, 1983; also see *CJRB*, May 26, 1983.
5. See speech by Zhou Taihe, *FBIS*, July 3, 1984, p. Pl.
6. Li Chong-huai, *"Liang tong" qi fei* [Take off on the "Two *Tong*'s"], (Wuhan: Wuhan University Press, 1986), pp. 47-48.
7. Xu Rongan, "The Central China Economic Region Must be Built," *Jianghan luntan* [Jianghan Forum] (hereafter, *JHLT*), Wuhan, No. 3, 1986), p. 8.
8. Zhou Taine, *op. cit.*, p. P2.
9. *FBIS*, February 5, 1986, p. K7.
10. Wang Mingquan, "Review and Prospects for Financial Reform," *Xuexi yu Shijian* [Study and Practice] (hereafter *XXYSJ*), Wuhan, No. 1, 1986, p. 6, and *CJRB*, December 22, 1986, and in *FBIS*, January 7, 1987, p. K22. These include setting up capital markets and experimenting with different forms of banking. Some of this is explained in a later section.
11. Interview with local urban economist, July 1, 1987.
12. The *People's Daily* warned of contradictions between the *jihua danlie* cities and their provinces on April 4, 1985, p. 5. As Vice Minister of the State Commission for Reform of the Economic System Zhou Taihe admonished, "Separate listing in the state plan for Wuhan does not mean creating a province within a province...organs at the two levels [i.e., provincial and municipal] should not sing their own tune or a tune different from that of Beijing."

13. *RMRB*, September 21, 1985, p. 1.
14. *FBIS*, October 17, 1984, p. K18.
15. A statement that the big contradiction between the supply and demand for capital had not yet been resolved appears in Wang Mingquan, *loc. cit.*
16. Wu Xinmu, Professor, Economics Department, Wuhan University on June 27, and an official local economist, *loc. cit.*, explained the new financial arrangements to me. Wuhan, Professor Wu said, has a formal retention rate of 14 percent, but it may keep some extra funds for such things as wage adjustment, flood control and other emergencies, bringing Wuhan's actual retention up to 20 percent of its receipts. Other cities' formal rates, again drawing on my interview with Professor Wu, are as follows: Shanghai, 18 percent; Tianjin, 21; Guangzhou, 24; and Shenyang, 23. According to the Municipal Planning Commission (June 23 interview), about 38 percent of Wuhan's income now goes to Hubei, whereas about 41 percent did in the early 1980s.
17. See the article on the reform in *Shehui kexue dongtai* [Social Science Trends], Wuhan, No. 17, June 10, 1984, p. 18; also, Li Chong-huai, *op. cit.*, p. 47.
18. Economist, *loc. cit.*
19. *Ibid.*
20. Wu Xinmu, *loc. cit.*
21. Interview with Yu Zhi'an at the Yangtze Energy Corporation, June 21, 1987. This supposed differentiation between economic and administrative powers was also alluded to by speakers from the City Planning Commission on June 23, and from the Wuhan branch of the People's Bank on June 24.
22. City Planning Commission, *op. cit.*; July 1, with official from Mayor's Consultative Committee; and June 22, with city office of the Industrial and Commercial Administrative Management Bureau.
23. People's Bank, *loc. cit.*
24. *Joint Publications Research Service-China East Asia* (hereafter *JPRS-CEA*)-*85-051*, June 5, 1985, p. 45, translated from *Jingji Guanli* [Economic Management] (hereafter *JJGL*), No. 2, 1985.
25. City Commission on Reform of the Economic System, interview with officials, June 22.
26. Economist, *loc. cit.* From what he said, it seems that the main problem now is that the city is unable to guarantee all the

material supplies for the firms, so that some retain their ties to upper levels, whether the province or, more often, the central ministry is ultimately in charge of them.
27. Interview at Wuzhong, June 29.
28. *Ibid*. The amount of supplies it obtained through the plan also decreased each year after this arrangement was installed—40 percent of total needs was met in 1983, but less thereafter, in accord with the amount of supplies the ministry had on hand at any time. Despite about an hour of questioning on this point, I couldn't clarify to my satisfaction just what the "base, guaranteed amount" really signified, given this decrease in amount supplied over time. But the important point for our purpose at this stage is that the plant's connection with Wuhan underwent no change with regard to its official supply source, despite the *xiafang*. Its revenues went entirely to Wuhan, but, as explained above, Wuhan had to turn over to the center the great bulk of these.
29. Yu Zhi'an, *loc. cit.*
30. Wu Xinmu, *loc. cit.*
31. City Planning Commission, *loc. cit.*; and economist, *loc. cit.*
32. *China Daily*, December 2, 1985, p. 1; "Record of Major Events in Wuhan's 1985 Economic System Reform," *XXYSJ, op. cit.*, p. 71; interview with MPC, *loc. cit.*; and economist, *loc. cit.*
33. Economist, *loc. cit.*
34. Interview on November 30, 1984, at the Wuhan Valve Works.
35. *XXYSJ*, No. 1, 1986, p. 71.
36. Li Chong-huai, "Prospects for Wuhan's Economic System Reform," *XXYSJ, op. cit.*, p. 4.
37. Economist, *loc. cit.*
38. The national press is constantly replete with admonitions to bureaus not to interfere with the new management powers for the firms. This has been going on since the initiation of the reforms.
39. *RMRB* Overseas Edition, November 6, 1986, p. 1. Also see Ta-kuang Chang, "The East is Red," *China Business Review*, March-April 1987, pp. 42-45, and article by Nina MacPherson in *South China Morning Post*, December 12, 1986, p. 1, reprinted in *FBIS*, December 12, 1986, pp. P3-4.
40. The law appears in *Guowuyuan gongbao* [State Council Bulletin], No. 521, December 20, 1986, No. 33, pp. 979-985, under the name "Enterprise Bankruptcy Law of the People's Republic of China"

(Draft).
41. *FBIS*, December 12, 1986, p. P4.
42. City Planning Commission, *loc. cit.*
43. As Yangtze Energy Corporation Manager Yu Zhi'an, *loc. cit.*, explained, the whole point of bankruptcy regulations is to motivate staff and workers to mobilize themselves to work harder under threat of failure. Consequently, it is to be expected that propaganda surrounding the radio plant's recovery would stress the role of the independent efforts of those working at the firm.
44. Reportedly, "For a long time, one third of the small state-owned enterprises in Wuhan have been running in the red or just breaking even." See *FBIS*, November 26, 1986, p. P7.
45. *CJRB*, June 9, 1987, p. 1. See also *RMRB*, July 7, 1987, p. 2, in *FBIS*, July 17, 1987, p. K6, which discusses "big fish taking care of little fish" and advanced enterprises contracting for the management of backward enterprises, both examples of one plant rescuing another one.
46. Interview at the Wuhan Chemical Machinery Factory, June 25.
47. The state-owned Wuhan Automobile Engine Plant, with 1,500 workers, was leased to 21 of its workers to manage in late 1986, "as the plant was on the brink of bankruptcy," according to the press release. This was reported in *Zhongguo xinwen she* [Chinese News], Hong Kong, November 20, 1986, and translated in *FBIS*, November 26, 1986, p. P7.
48. *RMRB*, May 24, 1987, p. 2.
49. City Commission on Reform of the Economic System, *loc. cit.*
50. A clear discussion of the various forms of the leasing system is by Zheng Li, Liu Zhao, and Xiao Wentong, "Different Forms of Rental Management," in *Honqqi* [Red Flag], No. 12, 1987, pp. 17-18.
51. *RMRB*, February 6, 1987, p. 5, translated in *FBIS*, February 10, 1987, p. K7.
52. *CJRB*, April 27, 1987, p. 1.
53. Discussion of the various forms of this system are recounted in Lu Dong, "Contract Management is an Effective Avenue for Invigorating Large and Medium-Sized Enterprises," *Honqqi* [Red Flag], No. 9, 1987, pp. 21-24 and Feng Baoxing, "Explorations of Some Questions about the Contract Management Responsibility System," *Honqqi* [Red Flag], No. 10, 1987, p. 25.
54. Wuhan had 44 experimental keypoints practicing this system as

of mid-1987, and intended to expand this to another 56 soon thereafter. This information was given to me at the Wuhan General Machinery Plant on June 24.
55. This was the case at the Central-South Large Commercial Building, as explained at an interview on June 29.
56. *Guangming ribao* [Bright Daily], July 18, 1987, p. 3, in *FBIS*, July 31, 1987, p. K9.
57. A *Beijing Review* article on the contract responsibility system in issue No. 34, August 24, 1987, p. 5 states that, "No longer does the enterprise carry out production in accord with the state's mandatory plan."
58. Wuzhong, *loc. cit.*
59. *Ibid.* Other interviews on leasing and on the management responsibility system were conducted at the Wuhan General Machinery Plant, *loc. cit.* the Wuhan Chemical Machinery Factory, *loc. cit.*, the Wuhan No. 1 Textile Mill (June 26), and at the Central-South Large Commercial Building, *loc. cit.*
60. Wuhan General Machinery Plant, *loc. cit.*
61. *Ibid.*
62. *RMRB*, July 24, 1987, p. 2, in *FBIS*, July 31, 1987, p. K5.
63. The Wuhan General Machinery Plant still gets more than half its steel from one Machine-Building Bureau; the Wuhan No. 1 Textile Mill gets 95 percent of its cotton from administrative channels, and the Wuhan No. 3 Printing and Dyeing Plant depends on its bureau for half of its cloth, but for more when supplies are tense, since the bureau knows just which textile mills under its control can fill the gap. The Bureau also helps this plant develop connections with the departments supplying electricity, "because we don't have direct relations with them." For a more thorough treatment of these issues of bureau assistance to and interference in the affairs of the firms in the reform period, see Andrew G. Walder, "The Informal Dimension of Enterprise Financial Reforms," in US Congress, Joint Economic Committee, *China's Economy Looks Toward the Year 2000: Volume 1. The Four Modernizations* (Washington, D.C.: Government Printing Office, 1986), pp. 630-645.
64. Mayor's Consultative Committee, *loc. cit.* On that day, the companies that stand between the bureaus and the firms were no longer to receive the management fees. At some point in the near future, these fees will no longer go to the bureaus either. At that point, the Ministry of Finance will have to give larger

financial allocations to the bureaus, since that will then become the bureaus' sole source of funding. The reasoning in reform circles on this point goes that once bureaus stop intervening, production will increase as firm incentives increase, and as a result the Finance Ministry will take in more taxes, giving it more funds to disburse to the bureaus.

65. Hu Jizhi, "Some Issues in Opening Wuhan's Financial Markets," *JHLT* No. 10, 1985, p. 14.
66. Li Chong-huai, "*Liang tong*," p. 62; *JPRS-CEA-85-008*, January 26, 1985, pp. 68-69.
67. *FBIS*, March 26, 1987, p. K40. The other four initial cities were Shenyang, Changzhou, Chongqing, and Guangzhou.
68. *FBIS*, July 23, 1984, p. P3; and Hu Jizhi, *op. cit.*, p. 17.
69. Wang Mingquan, *loc. cit.*, p. 6.
70. People's Bank, *loc. cit.*, p. 17.
71. Hu Jizhi, *loc. cit.*
72. *RMRB*, April 5, 1987, p. 3, in *FBIS*, April 7, 1987, pp. K29-33.
73. *RMRB*, April 7, 1987, p. 3.
74. *CJRB*, May 25, 1987, p. 1.
75. For Shenyang, see Hao Yanni, "Investigation into Strengthening Control over the Bonds Market," *JJGL*, No. 12, 1986, p. 30. For Shanghai's, see the *Los Angeles Times* April 27, 1987.
76. The *Changjiang Daily* used the term Wuhan Trust Investment Company (*xintuo touzi gongsi*) in the article cited here. Is this the same organ as the Financial Trust Company referred to above and also mentioned in my June 24 interview with a representative of the research office of the Wuhan branch of the People's Bank? I did not think to ask when I had the chance, but it seems that it must be.
77. *CJRB*, June 27, 1987, p. 1.
78. Specialized banks such as the Industrial and Commercial Bank by all accounts have little independence yet from the central, People's Bank.
79. For instance, Susan Shirk related to me that the prices at which given stocks can be sold is limited by regulations.
80. Economist, *loc. cit.*
81. *Ibid.*
82. *CJRB*, April 6, 1987, p. 1, and also interview at the Wuhan Bazaar, June 23.
83. Wuhan No. 2 Bicycle Factory interview, June 25.
84. Hu Jizhi, *op. cit.*, p. 15.

85. As explained by the Financial Research Institute of the Wuhan branch of the People's Bank, *loc. cit.*
86. Economist, *loc. cit.*
87. Qi Caozu, "Wuhan jingji tizhi zonghe gaige di huigu yu sikao," [Looking Back and Reflections on the Comprehensive Reform of Wuhan's Economic System] *Wuhan jingji yanjiu* [Wuhan Economic Research], Wuhan, No. 3, 1987, p. 8.
88. *CJRB*, April 11, 1987.
89. People's Bank, *loc. cit.* On the 27-city network, see *RMRB Overseas Edition*, October 11, 1986, p. 1; on the 55-city one, see *FBIS*, April 20, 1987, p. P3 and *CJRB*, April 11, 1987, p.1.
90. See *supra* note 23.
91. July 1 Interview and economist, *loc. cit.* This data is consistent with a report in *Liaowang* from July, 1985, translated in *FBIS*, August 1, 1985, which states that, "finance is still vertically and rigidly controlled by the higher levels. As capital is not circulating, it is very difficult to invigorate enterprises and cities."
92. *RMRB*, June 3, 1987, p. 2.
93. All of the following information comes from interviews with the firms cited: Wuzhong, No. 2 Bicycle Factory, *loc. cit;.* Wuhan No. 3 Printing and Dyeing Factory *loc. cit.*; Wuhan General Machinery Plant, *loc. cit.*
94. Enterprise groups are discussed in a commentator's article in the *People's Daily*, July 21, 1987, p. 2, translated in *FBIS*, July 29, 1987, p. K14. According to *People's Daily*, August 10, 1987, p. 2, there were then over 1,000 such groups registered.
95. *CJRB*, April 10, 1987, p. 1.
96. Xinhua, March 11, 1985, in *JPRS-CEA-85-034*, April 5, 1985, p. 72.
97. *RMRB*, May 8, 1987, p. 5, in *FBIS*, May 20, 1987, p. K14.
98. *FBIS*, February 5, 1986, p. K7.
99. Qi Caozu, *loc. cit.*
100. Wu Xinmu, *loc. cit.*, p. 8.
101. *RMRB*, August 10, 1987, p. 2.
102. *JPRS-CEA-85-035*, April 15, 1985, pp. 40-46.
103. *China Daily*, May 12, 1986; see also *FBIS*, July 29, 1987, p. K15.
104. *Beijing Review*, No. 29, July 22, 1985, p. 27.
105. Economist, *loc. cit.*; and Mayor's Consultative Committee, *loc. cit.*
106. *XXYSJ*, *op. cit.*, p. 70. I requested an interview with this committee, but was told that speaking with its members would duplicate material I got elsewhere. Probably I could have

collected more statistics had I had the chance to hold that interview.
107. Mayor's Consultative Committee, *loc. cit.*
108. City Commission on Reform of the Economic System, *loc. cit.*
109. Mayor's Consultative Committee, *loc. cit.*
110. Wuhan Chemical Machinery Factory, *loc. cit.*
111. Wuzhong, *loc. cit.*
112. On the increased role of the materials bureaus at all administrative levels under the reforms, see the article by Christine P.W. Wong in this volume. Her information comes from a pamphlet published in 1984 by the State Materials Bureau.
113. *RMRB*, August 31, 1987, p. 2.
114. The following data on information flows, the *caigouyuan* and the *xinxiwang* come from an interview at the Wuhan Chemical Machinery Plant, *loc. cit.* It fits with the analysis by Jean C. Oi, "Commercializing China's Rural Cadres," *Problems of Communism*, September-October, 1986, pp. 1-15, especially pp. 9-10, in which she speaks of Chinese peasants' difficulties in the new rural markets because of their lack of information about market demand and prices, and their heavy dependence on their local cadres for this. Also, N.A., *Wuhan Nianjian 1986* (Hankow: Wuhan Nianjian Bianzuan Weiyuanhui Bianji [Wuhan Yearbook Compilation Committee], 1986), pp. 356-359 outlines and describes the functions of the massive local information network created by government organs in recent years in Wuhan and indeed throughout the country.
115. Central-South Large Commercial Building, *loc. cit.*
116. Wuhan General Machinery Plant, *loc. cit.*
117. Wuhan Bazaar, *loc. cit.*
118. Wuzhong, *loc. cit.*
119. Economist, *loc. cit.*
120. Carl Riskin, *China's Political Economy: The Quest for Development Since 1949* (New York: Oxford University Press, 1987), p. 358. The first of these opened in Chongqing in 1984, according to *RMRB*, August 9, 1987, p. 2.
121. *JPRS-CEA-85-037*, April 17, 1985, p. 47, and *China Daily*, January 20, 1986, p. 1.
122. Deng Shaoying, "The Conditions for Development of Trade Centers and Their Functions," *JHLT*, No. 11, 1985, p. 14.
123. *China Daily*, March 3, 1986, p. 2. Barry Naughton, however,

informed me that for all these materials except steel, these figures probably represent closer to one-third of those in use.
124. Zhang Baohua, Dai Guanlai and Li Jiqing, "Countermeasures to Solve the Large Degree of Inflation in Prices of the Means of Production Outside the Plan," *JJGL*, No. 1, 1986. Yet another source states that in 1985, "Less than 20 percent of the steel products in the country circulated among enterprises and on the market." This is from Zhou Shoulian, "On the Means of Production Market," in *RMRB*, February 20, 1987, p. 5, translated in *FBIS*, February 27, 1987, p. K33.
125. *RMRB*, May 21, 1985, p. 1.
126. *JPRS-CEA-85-038*, April 22, 1985, p. 31. It is not clear what is being counted here or how. For *China Daily*, March 3, 1986, p. 2, states that "China now has 32,000 trading centers engaged in materials sales, 6,800 more than in 1980. If that's so these larger figures must pertain to markets of all sizes, the 130 noted above being only the most major ones. And *RMRB*, August 9, 1987, p. 2, cites 930 "urban industrial products trade centers."
127. Qi Caozu, *op. cit.*, p. 7.
128. City Planning Commission, *loc. cit.*
129. *FBIS*, January 9, 1987, p. K41. Wuhan's was actually set up on February 20, according to *CJRB*, May 12, 1987, p. 1.
130. Wuzhong, *loc. cit.*
131. *RMRB*, May 9, 1987, p. 2.
132. City Commission on Reform of the Economic System, *loc. cit.*, Wuhan General Machinery Plant, *loc. cit.*
133. Lin Ling, "Give Play to the Central Role of Cities, Develop Horizontal Economic Links," *JJGL*, No. 2, 1985, pp. 3-7, in *JPRS-CEA-85-051*, June 5, 1985, p. 47.
134. *JPRS-CEA-85-032*, April 1, 1985, p. 63.
135. I was told that the consolidation was done to eliminate those without experience, with insufficient capital and equipment, or without skill in business. Probably some were also operating in illicit ways. City Commission on Reform of the Economic System, *loc. cit.*
136. This analysis, a very common one, comes from Liu Guangjie, "Discussing the Position of 'Invigorating Circulation' in the Comprehensive Reform of the Urban Economic System," *JHLT*, No. 11, 1984, p. 7.
137. Deng Shaoying, *op. cit.*, p. 12, and City Commission on Reform of the Economic System, *loc. cit.*

138. *Ibid.*, pp. 14-17, and City Commission on Reform of the Economic System, *loc. cit.*
139. This was the remark of a management cadre at the Central-South Large Commercial Building, *loc. cit.* This is especially the case for the "specialized" centers that deal in one or several related products. The other type of center is "comprehensive."
140. This was the figure supplied by the City Commission on Reform of the Economic System.
141. *Beijing Review*, No. 28, July 15, 1985, p. 24.
142. The story, in April 30's *CJRB*, p. 1, says that the traders are from Zhejiang, Jiangsu, Guangzhou, Hunan and Henan, as well as other Hubei cities and counties. A cadre from the city's Industrial and Commercial Administrative Management Bureau said, *loc. cit.*, that individual firms from all over the country (28 provinces) come to Yangtze to get goods which they take home and sell, and that the market has influence throughout the whole country.
143. See *supra* note 64.

NINE

THE POLITICAL ECONOMY OF REGIONAL REFORM
AN OVERVIEW

Victor C. Falkenheim, University of Toronto

Over the past decade, a central but elusive goal of China's economic reform has been the development of "natural" forms of interaction among economic units. Reform advocates have deemed such reforms essential to remedy what they perceive as the critical structural defect of the economy, the subordination of producing and marketing units to administrative authority.[1] This problem, traceable to the imposition in the 1950s of centralized Soviet economic structures on an under-developed "small producer" economy, has resisted easy solution. The primary remedy, attempted first in 1957, and again in the late 1960s and early 1970s, emphasized decentralizing economic authority to provincial and local levels. However, rather than enhance flexibility, these reforms led to new forms of rigidity and inefficiency. In the view of one senior economist, the main result of the reform was to replace wasteful inter-ministerial conflict with equally wasteful and costly inter-regional economic conflict, neither conducive to the "unified development" of the economy.[2]

In the 1980s, two sets of reforms sought to overcome this problem. The first has stressed the progressive marketization of the economy, reducing the scope of direct, mandatory planning. The second, more accommodative of China's political-administrative realities, has sought to simultaneously decentralize the economy while fostering new forms of regional coordination. It is these regional reforms that form the subject matter of this chapter.

REGIONAL REFORMS: AN OVERVIEW

A wide range of regional experiments have been undertaken in recent years in virtually every sector of the economy. In the transport sector, for example, the administration of the railways and internal waterways have been revamped to create new and enlarged supra-provincial entities. In the commercial sector, higher order regional marketing systems have been developed in such regional centres as Wuhan and Chongqing. Supra-provincial economic zones have been created in the Shanghai area and the Pearl River delta area to foster cooperative developmental links between enterprises. All of these have been supported by a national effort to promote "lateral economic cooperation" (*heng xiang lian xi*) as a crucial ingredient in the structural reform of the economy.

These experiments have a number of features in common. First, they diminish the economic authority of provincial and prefectural levels of administration. Second, they enhance the role of economically central industrial and port cities and regional marketing centers. They appear in part to constitute an effort to create territorial-administrative jurisdictions more closely aligned to "natural" economic activities.

The goal of this chapter will be to describe the simultaneous efforts to decentralize economic authority while fostering new forms of economic cooperation. It will begin with a discussion of regional development policy which sets the context within which both sets of reforms have been undertaken.[3] It will then focus on their inter-relationship which has shaped the implementation of regional reform since the early 1980s.

ECONOMIC REFORM AND REGIONAL DEVELOPMENT

Decentralization of economic authority to provincial and local levels was a major feature of the early post-Mao reforms. Unlike comparable Mao-era initiatives however, it was accompanied by a great deal more policy flexibility. During the Mao era, local discretion meant the freedom to promote local agricultural and industrial self-sufficiency. In the Deng era, the premises of regional development have been thoroughly revamped. As set forth in the regional development chapters of the 6th and 7th Five-Year Plans (1981-1985, 1986-1990), the emphasis is first and foremost on regional specialization on the basis of comparative advantage, abandoning the goal of local "self-reliance."[4]

The slogan which best captured the new emphasis was the injunction to provinces to bring into play their "strong points," and sidestep their "shortcomings." This approach would lead to "greater, faster, better, more economical" growth.[5] In considering how such an approach might be applied to capital construction, a *People's Daily* editorial called for tailoring construction plans to local conditions as leading to "better economic results" and constituting an important part of the road to Chinese-type modernization.[6] Subsequent press commentary spelled out potential industrial and agricultural applications of the principle. The provinces of Hebei, Shandong and Henan with long sunny periods were climatically better suited to cotton cultivation.

Further, their superior experience and technology as well as their proximity to well developed textile centers argued for shifting resources out of grain production where per acre yields were low, into cotton where yields were high. Conversely, a province like Heilongjiang with an acknowledged superior potential as a supplier of marketable commodity grain might better concentrate on building itself into a "modern agricultural base" supplying North China's food grain needs. Coastal provinces, it was further suggested, might emphasize gains from foreign trade, playing a catalytic role in organizing local production for export. In particular, key cities such as Shanghai, with a developed industrial and technological infrastructure, were to emerge as pacesetters. As one editorial put it, "a favorable situation will certainly prevail in China's modernization program with all areas of the country vying against one another in demonstrating their own strong points."[7]

In enjoining the regions to "avoid their weak points," the new policy explicitly rejected the previous emphasis on building "relatively complete" industrial systems in each province. The earlier goal of developing a sufficient number of core industries in each province to sustain local industrialization efforts, built around the "five industries" (e.g. steel, cement, agricultural, machinery, etc.), and the strongly emphasized goal of grain self-sufficiency built around the "grain first" policy were now held to have been wasteful, ecologically damaging and economically counterproductive, in some cases retarding growth rather than accelerating it. In particular, the effort to promote grain and steel as "key links" irrespective of local conditions and resources was under sharp attack for "turning strong points into weak points."[8] In an editorial summary, a *People's Daily* commentator concluded, "Thirty years of experience in construction have proved that if every area builds on the same economic patterns, it will hinder development of strong points and even turn them into weak points."⁹ In defending the new emphasis on regional specialization to skeptical local leaders anxious to defend their stake in existing industrial structures, Central officials made several points in justification. First, they argued that development strategies, meshed with local resources, would maximize local growth rates once the shackles of a uniform economic policy were thrown off and unprofitable industries closed down. Second, even if there were short-run negative consequences to the local economy, these had to be accepted in a spirit of sacrifice for the long-term common interest. Third, they criticized the view that all regions had to develop equivalent economic assets. As one commentary put it, "some comrades have mistaken the idea of 'balanced distribution' to understand it as the elimination of differences in levels of economic development." This interpretation was based on an incorrect understanding of Engels' notion of "balanced distribution" which called only for avoiding excessive concentration of large industries in a few big cities.[10]

Significantly, however, advocates of the new regional policy acknowledged directly that the new policies would in all likelihood increase relative disparities as better endowed and more productive regions outstripped their less-favored brethren. Commenting on the implications of the new economic role assigned to key cities, one article queried rhetorically, "Will bringing the role of key cities into full play result in more uneven economic development and slow economic growth in relatively backwards areas?" and answered in the affirmative. "We must recognize that the uneven development of various regions is an objective reality that cannot be overcome according to short term subjective desires." Drawing out further the distributive import of this policy another article noted "common prosperity" should not be equated with "simultaneous prosperity" or "equal prosperity" egalitarian notions which if applied would hinder development.¹¹

Yet, if the main thrust of policy has been away from an explicit redistributionist commitment in the direction of the most efficient deployment of resources and investment on a national scale, it has not meant an abandonment of the long-term goal of balanced development. This can be clearly seen in the chapter on regional development in the 6th Five-Year Plan which calls not only for "making the most of the coastal areas," but for accelerating the development of energy, communications, raw material, and semi-finished material industries in the inland areas:

> On the premise of fulfilling their quotas of cotton, wool, hemp, silk, sugar, and tobacco, the inland provinces and autonomous regions shall, in line with special features of their local resources, develop the production of consumer goods in a planned way so as to raise their degree of self-sufficiency in agricultural products for daily use, and maximize the inland's agricultural potential.[12]

Further, the plan pledged strengthening "planned industrial construction" in the minority nationality regions, albeit "in accord with local resources," and promised continued support in the "financial, material, and technological realms." To give concrete evidence of this commitment, the plan called for an annual 10 percent increase in the level of subsidies, and a special annual allocation of 500 million *yuan* for development assistance, in addition.[13] These policy emphases were made even more emphatic in Section III of the 7th Five-Year Plan which called for establishing "correct relations" among the principal regions of China by speeding development in the "East Coast Region," building energy and semi-finished materials industries in the "Central Region" and making active preparation for long-term development of the "Western Region."[14]

EMERGENCE OF ECONOMIC LOCALISM

Faced with policies which in some instances threatened long-cherished development schemes and in others opened significant new opportunities, provincial and local leaders responded vigorously. While there were few direct challenges to the newly enunciated principles, there was a good deal of special pleading framed in a fashion consistent with the new regional policy emphases. This pattern was clearest in the editorial responses of the underdeveloped minority nationality regions, which sought to turn the new slogans to regional advantage. For example, Inner Mongolian (Nei Mongol) writers agreed that past efforts to develop heavy industry in the region had often been costly and wasteful, and had resulted in a failure to tap the region's comparative advantage, "block(ing) up our route to wealth." Yet, after a glowing inventory of the region's rich natural resources, local leaders queried, "how to develop these?" Part of the answer, they conceded was "self-reliance" and "joint development with other fraternal provinces," but "active support from the State" was needed as well. "Banks," the writer suggested, "should give preferential treatment to Nei Mongol in loans."[15] The call for continued compensatory treatment was echoed in Yunnan commentaries. Yunnan, one writer noted, in a less-than-subtle reminder of the economic realities, had "over the past 30 years carried out the strategy of catching up with and surpassing other provinces in a bid to narrow the differences with the national average level and advanced region, but *these efforts have resulted in bigger and bigger differences*" (emphasis added).[16]

In the Northwest, where official policy called for deferring major development initiatives to the 21st century, and for holding off on major early infusions of capital, some provincial authorities called openly for more money to accelerate development. In the aftermath of the CCP's 12th Congress which proclaimed the goal of quadrupling national income by the year 2000, Ningxia authorities acclaimed the goal, citing rich opportunities for coal and hydro-power development. However, they noted pointedly, that despite progress since 1949, "the gap between Ningxia and the country has widened over the past 32 years," hardly befitting a national minority region with superior resources and potential. "For the sake of enabling the fraternal nationalities to overtake the advanced within a certain period and realize the Four Modernizations together, *it is necessary to have a higher rate of development than the whole country*" (emphasis added).[17]

A particularly interesting example of regional interest articulation can be seen in a Heilongjiang commentary which conceded that the province had many potential strong points as a raw material and food grain supplier which could be developed and endorsing the proposal that the province be built into a modern agricultural base with an emphasis on commodity grain production. But the analysis continued, "the key to bringing strong points into play lay with the leadership" and the degree to which leadership action was in conformity "with economic laws." In particular, the problem was the underpricing of coal. Citing senior economist Xue Muqiao's observation that Heilongjiang's "four trumps" were grain, timber, coal and petroleum, the article concluded that because of problems of price, those four trumps have not been brought into play, and the more they are, "the greater the loss suffered by the province." To add insult to injury, manufactured goods from other provinces were costly and yielded high profits to outside manufacturers. "It is hoped, that the central authorities will consider and solve the problem of exchange of means of production at unequal values," so as to "bring our province's strong points into play."[18]

If China's less developed areas were quick to assert their claims for central assistance in the development of their assets, other more developed regions were not tardy in challenging specific decisions which impinged on their interests. Not long after the 1979 State Council decision to emphasize Shanxi coal development, competing claims were made from other regions. While endorsing the merits of the Shanxi coal development program, one writer called for simultaneous efforts to "be made to actively develop coal mining in the five provinces of Shandong, Anhui, Henan, Guizhou, and Yunnan," which the article noted, could claim one-fifth of all China's known reserves. These reserves were also of high quality and easily exploitable. Moreover, they were closer to the southern markets where they were most needed.[19] Similar unhappiness was also registered in the capital city of Beijing over its new designation as a cultural and tourist center. An editorial in the *Beijing Daily* defending the "four point decision" on Beijing noted that guidelines for work in Beijing "should by no means be the same as those for industrial cities such as Shanghai, Tianjin, Shenyang, and Wuhan," and rebuked those who complained about the downgrading of Beijing's economic role. "We must learn," the editorial stressed, "to assess Beijing's role as capital from the country's overall situation and not confine ourselves to a single region and think only in terms of output."[20]

Beyond rhetorical responses many areas have simply chosen to ignore or claim exemption from the logic of specialization where its application imposed serious financial or other penalties. Provincial revenues, employment, foreign exchange holdings, etc. are all dependent on local industries, however inefficient. The long-term economic prospects for each region tend to be seen as contingent on upgrading inefficient local industries, not in shutting them down. Local resistance to central constraints became particularly evident in the course of economic "readjustment." Sichuan Province's response to calls for a shift away from heavy industry to light industrial production is quite typical:

> The heavy industry of Sichuan Province has great advantages and it has developed through many years of frugal living by people across the country. It is an invaluable source of wealth and therefore has to fully display its strong points. The *readjust-*

ment in the province must be carried out according to the reality of the province and we must in no way totally readjust the structure of heavy industry to that of light industry (emphasis added).[21]

A similar claim for exemption from central guidelines on the grounds of "local realities" was expressed by industrially lagging Fujian Province, whose governor in an unusually frank statement noted that the province's backwardness was a function of the past lack of "key projects" in turn reflecting strategic military considerations. Although "the country as a whole must close, suspend and merge enterprises," the governor said, "if we follow the same route without carrying out specific analysis, we are not seeking truth from facts." Fujian, he went on, could not develop "on the basis of tea production, carving, sugar and tea...but needs key industries," including oil refineries, chemical engineering, hydroelectric development, airports, railroads etc. He tactfully concluded, "Certainly, we must also promote light industry."[22]

Localities have been compelled in a number of instances to ignore the principle of "specialization and exchange" because of supply scarcities, a point frankly conceded by the leadership. "Some localities," one commentary noted, "have no choice but to carry out coal and iron mining because they are not supplied by other localities. Some localities are suited for growing economic crops, but if the state cannot guarantee a grain supply for them, they must sacrifice their strong points."[23]

On the policy level the call to "develop strong points" has led to vigorous inter-provincial competition for resources with provinces seeking to expand production, promote new products thus generating local revenue. Among the stratagems developed have been efforts to control local raw material production for local processing, the development of local preferential purchasing programs designed to protect infant regional industries, the development of local trademarks and local marketing operations both overseas and domestically, the development of provincial shipping fleets and provincially controlled port outlets to save service and commission charges, etc. and to sidestep constraints on local capital construction. While these tactics have sprung logically from the policy of urging provinces to build on their own strong points, which as one commentary noted "they tend to do in their own way,"[24] it was made possible by result of the dramatic increase in local extra-budgetary revenue as a consequence of local profit retention schemes, enormously inflating the discretionary resources in the hands of the localities. The consequences of this as well as the new degree of fiscal flexibility for efforts to control local capital construction were well captured in the following editorial calling for tighter control:

> Because the channels of funds are so numerous and measures for organizing an overall balance and exercising concrete control are lacking things remain in disarray. This finds expression chiefly in the fact that when internal funds decrease, foreign capital is available: when there are less financial funds, there are more bank loans, when a budget is cut, funds can be raised outside the budget; when investment in capital construction is reduced, funds can be obtained from the outlay for technical innovation measures. The saying is prevalent at the lower levels, "When we have no funds for capital construction we can obtain funds from the outlay for technical innovation measures. When we have no funds for technical innovation measures we can apply for bank loans. When the construction bank does not give us a loan we can ask for a loan from the People's Bank. When we cannot raise funds at home we can seek foreign capital."[25]

One of the more concrete inventories of resulting problems was offered at a national finance conference in 1982 by Premier Zhao Ziyang who noted that efforts to stimulate the economy and decentralize power "for the purpose of whipping up the enthusiasm of the local authorities" had numerous consequences which "affected our efforts to take the whole country and overall situation into consideration." Many localities "instead of surveying...problems on the basis of the overall interests of the state, serve their own interests, disobeying the unified plans of the state." Zhao denounced competition in the foreign trade and "blind local construction" adding that "in some localities the tasks for state monopoly and compulsory purchase have been shirked; and economic blockade between one district and another is fairly serious." Zhao cited as a specific example the overproduction of ferroalloy products for export in North and Northeast China which led not only to overstocking but to straining the limited energy resources of the region, depressing other areas of the economy.[26] *People's Daily*, calling for recentralizing of economic controls, targeted particularly the proliferating efforts of provinces to develop profitable and popular consumer good industries, "disregarding subjective and objective conditions." The editorial singled out the case of a "certain province that has built more than 40 washing machine factories in an unplanned way," noting in passing similar problems in the overproduction of radios, televisions, wrist watches, bicycles and electric fans. The same editorial inveighed against local protectionism:

> For the sake of promoting the sale of locally produced goods of poor quality, some districts have resorted to blockade measures to keep quality products of other provinces and municipalities from entering their domain. This will not be tolerated. No matter where they are located these districts are all part of the PRC. As long as they are PRC they can be sold in any part of the country....We must permit legitimate competition and not protect the backwards.[27]

RECENTRALIZATION

The response of central authority to this combination of advocacy and selective implementation has been mixed. For the most part, central planners have continued to insist on the regional division of labor incorporated into the

Five-Year Plans. In addition, they have been responsive to local pleading which seemed consistent with these principles. For example, the steady expansion of foreign trade rights, first to Guangdong and Fujian, then to Shanghai, Tianjin and Guangzhou, and finally to all 11 coastal provinces, was in all likelihood a response to local lobbying which was perceived as legitimate.

Beyond this, the Center has been sensitive to legitimate local equity and welfare concerns, providing for example in Chapter 19 of the 7th Five-Year Plan for special support to minority nationality areas, frontier areas, and poverty-stricken areas of the country. The Center has been careful to assuage concerns of interior areas arising out of the new emphasis of "key" and coastal cities as illustrated in Yao Yilin's assurance that promoting the role of key cities as "key link" should be done only after full consultation. It was necessary to safeguard the interests of the regions, by "soliciting and respecting the interests of provincial, municipal and autonomous regional governments," adding that "in view of the great disparities existing in our massive country...(central departments)...must avoid making unilateral and arbitrary decisions."[28] Similar emphases accompanied discussions of economic cooperation between the advanced regions and the lagging provinces, asserting that such cooperation was not intended solely to benefit the processing centers of the coast but rather was equally intended to assist the interior provinces, a process dubbed the "East-West dialogue." One article commenting on the massive gap between per capita income levels (GVIAO) in China's leading city Shanghai with over 5,500 *yuan* (1981) and last-ranking Guizhou with 303 *yuan* per capita (1981), noted that national economic development itself would be impeded "unless the differences between the developed and underdeveloped regions are narrowed through technical cooperation."[29]

Of greater concern, however, has been the loss of economic control, which has led to strong efforts at checking economic localism. Beyond ideological exhortation the government has sought to strengthen accounting and banking controls, refine construction authorization procedures to limit local industrial development which contravenes national guidelines. In the sphere of foreign trade, a host of new customs and regulations including direct import and export controls have sought to rein in the flexibility of the past few years. The use of treasury bonds to sop up excessive local liquidity and the imposition of surcharges for extra-plan capital construction are just a few of the additional measures being undertaken to reestablish coordinated growth strategies. In the planning and procurement sphere, tighter and more rigidly enforced quotas are being fixed to assure raw material supplies to the larger industrial centers. Defending this recentralization of economic authority, Hu Yaobang noted in his report to the 12th Congress of CCP that the "state must concentrate needed funds on key development projects in their order of importance," and "overcome undue decentralization in the use of funds." He admitted that competing locally funded projects "may seem badly needed from a local point of view" but they had to give way to "the idea of coordination of all the activities of the nation like moves on a chessboard."[30]

INTERREGIONAL ECONOMIC COOPERATION

This reimposition of central controls, while seen as necessary, was also viewed as a regressive interim measure on the road to more fundamental restructuring. From the early 1980s onwards, in fact, the need to free enterprises from both local and ministry/bureau controls was perceived as the key to basic economic reform. It is this latter concern that in a large part accounts for the growing interest in measures designed to promote "lateral economic cooperation" in the mid-1980s. Initially, such regional cooperation was conceived of as helping to "narrow the gap between advanced and less advanced" areas through cooperative ties in technology and production sharing, and in alleviating supply difficulties on an exchange or barter basis. As embodied in the 6th Five-Year Plan however, a more encompassing and formal notion of cooperation emerged, calling for the development "of material technologies and economic exchanges" based on contractual ties between enterprises and localities.[31] The reform thrust of this policy was given greater emphasis in the October 1984 10-Point Decision of the CCP Central Committee on Economic Reform which called for promoting lateral economic ties in the context of "smashing blockades" that stood in the way of a domestic "open door."[32] In the 7th Five-Year Plan lateral economic ties were given pride of place among restructuring reforms and were described as the key to "development of large scale socialized production" and "powerful rebuff to barriers between departments and regions."[33]

As an earlier editorial described the anticipated benefit of fostering such linkages:

> By opening ourselves up internally, lifting regional blockades and ending departmental fragmentation, we provide the necessary conditions for commodity comparison in terms of price and quality and for competition and the full operation of market forces.[34]

While a critical focus of lateral cooperation efforts has been to stress voluntary contractual combination among enterprises with "large and medium sized enterprises as the mainstay," the concrete experimentation with programs seeking to promote such ties has focused on the facilitating role of economic networks and economic zones frequently based on central cities. Thus, the economic cooperation section of the 7th Five-Year Plan, built on the three years of pilot programs in regional and city-based economic cooperation, has proposed to structure economic cooperation in terms of a three-tier network of economic zones.[35] In the first tier are large economic zones based on the largest cities and regions, including the Shanghai Economic Zone, the Northeast Economic Zone, the Shanxi Energy Zone, the Beijing-Tangshan-Tianjin Zone, etc. Below the first tier, a second echelon is to be established consisting of secondary economic zones linking provincial capitals and designated ports and cities along vital communication lines. Finally, a third echelon of zones is to be organized around cities directly under the jurisdiction of provincial governments. While some industry-specific conglomerates will transcend regions and be national in scope, e.g., the automotive, petrochemicals, and computer hardware industries, much of the cooperative activity undertaken will be shaped by regional and local structures.

What precise economic relationship will emerge remains to be defined. In many Chinese analyses, no distinction is drawn between "lateral economic links" and "inter-regional cooperation," and in a number of accounts they are explicitly equated.[37] In fact, the forms and content of cooperation subsumed by both are highly diverse. An inventory of inter-regional cooperative ties in North China published in early 1987 included among the 20,000 such initiatives the following: multi-province joint investment projects in resource development, transportation and manufacturing; coastal city investment projects in raw material production; industrial combines organized to increase production of high quality products; technology cooperation agreements; merchandising cooperatives, etc.[37] The sanctioned legal forms of such cooperative relationships include equity joint ventures, cooperative joint ventures, compensation trade agreements, as well as informal orderly marketing agreements among competing firms.

The scale of economic activity encompassed by such cooperative initiatives is hard to determine, but reported totals for the 6th Five-Year Plan period indicated that there were some 70,000 cooperative projects with a total investment of 34 billion *yuan*, with approximately two-thirds of the projects falling into the category of technical cooperation projects. A *China Daily* report, summarizing a 14-province survey of lateral ties, listed 36 transregional economic cooperation areas, while a JETRO analyst identified 24 relatively integrated economic cooperation zones and 100 looser regional economic cooperation zones had been set up as part of the effort to promote lateral links based on economic networks.[38] There appears to be significant overlap between inter-firm cooperative links and those involving extensive government-to-government relations at the local level and in fact, most accounts suggest a natural dynamic shaping the development of inter-regional cooperation. Wang Daohan, Mayor of Shanghai in describing the growth of Shanghai's cooperative economic links with other areas has described it in evolutionary terms: material exchanges and processing have grown into joint exploitation of natural resources and joint production; technical exchanges of a general nature have developed into well-planned and well-organized technical cooperation and transfer of technologies, study visits have been replaced by well-planned training programs and exchange of technical personnel. Cooperation between individual enterprises has expanded into joint economic cooperation with official backing.[39]

The fact that many of these cooperative relationships are mediated through and directly involve local governments as initiators and participants has implications for the larger goal of reform. Cooperation clearly increases flexibility and offers advantages to the contracting parties but it makes few inroads into the structural problems that lateral cooperation was intended to mitigate. Its main effect is to establish another arena within which inter-local competition can be carried on. The ability of regions to advance their own interest in setting the terms of cooperation depends on the kinds of economic and leverage and resources localities can wield. Clearly, the advanced coastal or urban industrial centers can play their strong suits which include their advanced technology, infrastructural cost advantages, and the not negligible factor of central policy support to assert their own interests. Tianjin's call for assured raw material supplies, or the right to freely substitute imports where supplies are lacking, and Shanghai's pleas for a clarification of the terms of technology transfer suggest the issues about which such competition now revolves. Resource rich provinces such as Anhui, Shanxi and Inner Mongolia are aggressively seeking investment, technology and a share of the processing as the price for security of supply. Although 90 percent of Shanxi's annual coal production of 100 million tons, for example, is assigned to the state, 10 percent is available for trading, and one illustrated use of these supplies is the technical assistance agreement with Zhejiang and Jiangsu to upgrade Shanxi's small chemical works in return for coal.[40]

The importance of these negotiated agreements to both parties can be seen from the case of Jiangsu which acquired between 20-30 percent of its requirements in cold-rolled steel and timber between 1980-1982 through cooperative arrangements.[41] One side effect of these negotiated cooperative links has been to make supplies more costly to the bigger centers. Certainly in the case of coal, the trade in short supplies has been driving the cost of coal up, alarming the State Economic Commission. "Not enough coal," complained Zhang Jingfu, "is under unified distribution. Many places have carried out coal production through coordination," with the result that the price "is constantly increasing."[42] The lingering local affection for centrally distributed, assured supplies comes through clearly in statements on economic cooperation by Tianjin Mayor Li Ruihan, who stressed that the "guiding ideology" shaping such cooperative ties should be that "regional coordination should be guided by the state plan." In contrast, a certain relish for informality appears evident in the comments of Inner Mongolian First Party Secretary Zhou Hui: "We can see the necessity for strengthening the plans and guidance for conduction of economic and technical cooperation, but the plans for cooperation differ from assigned state plans" and "make up for the deficiencies of state plans."[43]

These conflicts are at the heart of the bargaining process that shapes the substance of economic cooperation. As one conference put it in considering obstacles to cooperation in the Shanghai Economic Zone:

> At the first mention of integration, everyone will scramble to be the head of the dragon. No one wants to be its body or tail. But there is only one dragon. Given a head but no body no tail and no claws how can a dragon exist?[44]

Similar jurisdictional conflicts over the location of canning and pulp residue paper industries has plagued effort to rationalize regional development strategy in the Minnan Economic Zone. Because of these problems there has been disagreement over the extent to which central intervention is necessary to foster such cooperation even at the risk of compromising the principle of "voluntariness." Thus when Yao Yilin was urging the need for a fully consultative process in achieving regional and interregional cooperation,[45] an article in *People's Daily* of the same period was arguing that "in forming joint enterprises, it is not enough to stress voluntariness only, it is also necessary to push this work from the higher level to the grass roots, to take economic and administrative measures not to let things drift." "Administrative intervention," the article asserted, was necessary to solve "the knotty problem of 'ownership by region'" and had to be seen as "necessary scientific administrative intervention" which did not run counter to "economic laws."[46] Zhao Ziyang in discussing the problems inhibiting regional and inter-regional integration noted that the "key issue is profits," and that only when deadlock resulted was "the State Council and the State Planning Commission to be called in" to resolve and mediate them. And many of these problems go beyond conflicts of interests. An analysis by a director of a State Council research office pointed to a crucial problem in noting that the continued obligation of localities to fulfill mandatory plans compelled them "to tightly control the economic activities within their jurisdiction," even if "inevitable restrain enterprises from developing lateral economic ties." The analysis continued by noting that the problem increased in 1984 and 1985 when raw materials were in short supply and prices rising leading to localities "protecting their own interests by restricting or banning movement of production materials and consumer goods," even "setting up checkpoints along their borders."[47]

It was clearly difficulties of this kind that led Tianjin Mayor Li Rui-huan to call for policy solutions to overcome the widespread view that establishing cooperative ventures is "a hard and thankless job." Li noted:

> The importance of cooperation is recognized only when something is desperately needed. When it comes to developing joint operations or transferring products and technologies, people are full of misgivings. They are afraid that things might not work out to their advantage if they lose their superiority. They think that cooperation is all right as long as they are the receiver and not the giver.[48]

To address these Li called for incorporating cooperative undertakings into the planning process, allowing them access to bank loans, the flexibility to adjust intra-firm price and profit differentials, and to establish separate targets for cooperating entities so that credit for projects is properly assigned.

THE ROLE OF "CORE" CITIES

The fact that these forms of cooperation have been slow to materialize and have barely dampened down the inter-regional rivalry they were meant to overcome has led to continued interest in enhancing the economic role of "core" cities. Advocates of such a restructuring have, from the outset, seen "central cities" as "replacing the existing system of provincial administration" in effect restoring the historic roles of such cities as Shanghai, Tianjin and Guangzhou as regional centers linked in a "nationwide economic network."[49]

The economic rationale for such proposals is premised on the view that "economic networks" are the natural result of the development of the commodity economy and with cities as their core play a key role in regional division of labor in a developing economy. A similar imperative operates within the context of planned-economy, advocates contend, and the proper reliance on "networks" will promote rationalization and growth within a planned economy. As one commentary put it:

> only by regarding key cities as our bases can we form the technological development networks which integrate our coastal areas with the hinterland, link the internal with the external and promoted...coordinated development.[50]

To do so the leadership has to overcome the tendency to treat key cities as "purely administrative areas," overlooking the fact that "they are key regional or even national economic centers."

This urban-focused reform strategy not unnaturally was embraced enthusiastically by city fathers across the country eager to give substance to their own "key" roles. In this process a variety of new issues arose, illustrating the pitfalls of partial reform, problems frankly acknowledged by advocates:

> [The] question that worries all of us is that of whether, in bringing into play the role of key cities, cities will create new barriers....Quite a few cities want to set up their own systems by putting themselves at the core. Some of them do not want to integrate themselves with other places and some sever their relations of coordination with other places. These practices cannot correctly bring into play the role of cities.[51]

In summarizing some of the lessons derived from 52 pilot reforms examining the ways in which cities could be used to promote lateral economic ties, an analysis in *Jingji guanli* noted that "smashing blockades," was a "more complicated reform" than separating the economy from administration. The key was "material force," with cities best served by relying on good products at cheap prices to "smash blockades." One way to ensure this is to rely on enterprises to take the initiative. "What is called giving play to the central role of cities cannot be accomplished by city governments...but mainly needs to give play to the role of enterprises."[52] Still other commentaries warned against devolving authority over enterprises to the cities lest a situation like the "warring states" develop with China "carved up into hundreds of natural economic zones."[53] Calling for genuinely voluntary economic combinations, a *Renmin ribao* writer noted that "some central cities tightly bind the enterprises within their administrative jurisdictions under the excuse of giving play to their central role, thus forming new blocs." We cannot "regard economic associations arranged arbitrarily by administrative institutions as horizontal economic associations."[54]

Clearly this tendency to compartmentalization which belies the optimistic view that "unlike other localities," cities are "disinclined to close their doors," and hence can be relied on to "unclog circulation," arises from the structural contradictions of a system in transition.[55] Many of the solutions being offered which include further reform of the tax, finance and planning systems should alleviate some of these problems. Others which impose tighter central scrutiny and approval procedures for "lateral combinations" will also address short-run problems in the regional reform. It has been suggested by some that fostering lateral economic ties may be the pivotal structural change in "an urban economic reform...looking for its key structure."[56] If so, it will be a long and difficult road to its full realization.

NOTES

1. Hu Qiaomu, "Act in Accordance with Economic Laws," *NCNA*, October 5, 1978, in *FBIS*, October 11, 1978, p. 12.
2. *Ibid.*, p. 9.
3. *Renmin ribao*, April 13, 1984, in *FBIS*, April 26, 1984, pp. K10-14.
4. *Beijing Review*, No. 22, May 30, 1985, Supplement, pp. VI-VII.
5. *FBIS, August 5, 1980, L15.*
6. *FBIS, June 19, 1980, L3.*
7. *FBIS,* July 15, 1980, L14-15.
8. *FBIS,* May 20, 1980, L8.
9. *FBIS,* July 15, 1980, L13.
10. *FBIS,* September 18, 1980, L11.
11. *FBIS,* June 19, 1980, L4.
12. *Beijing Review, op. cit.*, p. VIII.
13. *FBIS,* May 28, 1980, R6.
14. *FBIS,* April 18, 1986, K21.
15. See *supra* note 13.
16. *Sixiang zhan xian* (Kumming) 5, pp. 14-16 (October 24, 1982).
17. *Ningxia ribao*, October 15, 1982.
18. *FBIS,* September 15, 1980, L6-7
19. *FBIS,* August 22, 1979, L10.
20. *FBIS,* June 30, 1980, R2.
21. *FBIS,* October 20, 1981, O4.
22. *FBIS,* April 16, 1982, O3.
23. *FBIS,* August 5, 1980, L8.
24. *FBIS,* January 5, 1981, L16.
25. *Ibid.*
26. *FBIS,* April 1, 1982, K1-2.
27. *FBIS,* February 26, 1982, K2.
28. *FBIS,* April 27, 1981, K2.
29. *Beijing Review*, Vol. 33, August 6, 1982.
30. See *supra* note 12.
32. Chao Yu-sen, "Promotion of Lateral Economic Ties on Chinese Mainland," *Issues and Studies*, Vol. 22, No. 7 July 1986, p. 5.
33. *Xinhua*, March 23, 1986; in *FBIS*, April 1, 1986, K1.
34. JPRS, *China: Economic Affairs (CEA)*, 102, 1985, p.62.
35. *FBIS,* April 18, 1986, K2.
36. *Beijing Review*, Vol. 30, No. 7, February 16, 1987.

37. *Ibid.*
38. *FBIS*, April 17, 1986, K7: *China Daily*, July 24, 1986; Masahuru Nishida, "Recent Moves Towards Regional Authority," *China Newsletter*, No. 68 May-June 1987, pp. 16-17.
39. *FBIS*, January 28, 1983, K10-11.
40. *FBIS*, April 4, 1984, K14-15.
41. *FBIS*, November 12, 1981; *FBIS*, May 3, 1983, T4.
42. *Ibid.*
43. Radio Hohhot, February 27, 1983, in *FBIS*, March 10, 1983, K4.
44. *Shijie jingji daobao*, April 16, 1984, p. 1, in *FBIS*, May 14, 1984, O2.
45. Xinhua, April 25, 1981, in *FBIS*, April 27, 1981.
46. *Renmin ribao*, March 31, 1981, in *FBIS*, April 1, 1981, K5.
47. *Renmin ribao*, April 7, 1986; in *FBIS*, April 17, 1986, K8.
48. *Jingji guanli*, No. 1, January 5, 1985, in JPRS-CEA-85-O35, p. 40.
49. *Qiye guanli*, No. 3, 1981, in *Beijing Review*, No. 37 September 14, 1981.
50. *Tianjin ribao*, July 27, 1982, in *China Report: Economic Affairs, (CEA)* No. 263, p. 15.
51. *Renmin ribao*, October 21, 1983, p. 5, in *China Report: Economic Affairs*, No. 400, p. 92.
52. *Jingji Guanli*, no. 2; in *JPRS-CEA-85-O51*, 85:O51, p. 49.
53. *Jingji Guanli*, November 5, 1983; in *JPRS-CREA-84-O12, p. 21.*
54. *Renmin ribao*, May 2, 1986; in *FBIS*, May 9, 1986, K4-5.
55. *Renmin ribao*, October 21, 1983, p. 5; in *China Report: Economic Affairs* no. 400, p. 91.
56. "New Ways for the Regional Economy: Lateral Links," *China News Analysis*, No. 1311, June 1, 1986, p. 2.

EDITORS

Ilpyong J. Kim is professor of political science at the University of Connecticut. Dr. Kim has written and edited six books and published many articles in professional and academic journals on comparative and international politics of East Asia: (China, Japan, and Korea.) He received his Ph.D. from Columbia University, and has taught in the United States and Japan.

Bruce L. Reynolds has been associate professor of economics at Union College since 1974. He has travelled extensively in the People's Republic of China and was an instructor in western languages and literature at Tunghai University in Taichung, Taiwan from 1966–1969. Professor Reynolds received his Ph.D. in economics from the University of Michigan. He is the author of *Reform in China* and *The Chinese Economy in 1980*.

CONTRIBUTORS

Robert F. Dernberger is a professor of economics at the University of Michigan, Ann Arbor. He received his Ph.D. from Harvard. He served as special editor of the "Asia Pacific Report" (1989), an annual report on Asia published by the East-West Center in Honolulu, Hawaii. This was a special issue devoted to recent developments in China. He is a past-president of the Association of Comparative Economic Studies. His publications include *China's Development Experience in Comparative Perspective,* and with Allen S. Whiting, *China's Future.*

Victor C. Falkenheim received his Ph.D. in political science from Columbia University. He is presently chairman of the department of East Asian studies at the University of Toronto, a member of the review panel of the Committee on Scholarly Communication with the People's Republic of China, National Academy of Sciences, Washington D.C., and the Editorial Advisory Committee of Pacific Affairs. His publications include chapters in *Citizens and Groups in Chinese Politics, China in the 1980s: Reforms and Their Implications,* and numerous journal articles and papers.

Thomas Baron Gold is assistant professor of sociology at the University of California, Berkeley and serves as interpreter-escort with the National Committee on US-China Relations, Inc., and the Department of State. He has accompanied Chinese academic, cultural, political, and sports delegations to the US, and American cultural and academic delegations to China. He was formerly a staff associate of the Committee on Scholarly Communication with the People's Republic of China and instructor in the department of foreign languages at Tunghai

University, Taichung, Taiwan. He received his Ph.D. in sociology from Harvard University.

Justin Yifu Lin is a professor of economics at Peking University in Beijing. He received his MA in Marxist economics at Peking University, Beijing, and his Ph.D. in economics at the University of Chicago. He is editor of Modern Economic Series (Chinese, published in Beijing,) and was a postdoctoral fellow at the Economic Growth Center, Yale University, and a consultant on agricultural and rural development at the World Bank. He is presently teaching at Peking University and engaged in research at the Economic Growth Center at Yale University.

Ramon Hawley Myers is curator-scholar of the East Asian Collection and senior fellow in the Hoover Institution as well as adjunct professor at Stanford University. He received his Ph.D. in economics from the University of Washington (Seattle). He served as a Fulbright lecturer in the department of economics at the Chung-chi College at the Chinese University of Hong Kong and was a visiting professor at National Sun Yat-sen University, Kaohsiung, Taiwan. His research has focused on Chinese economics and social history and American foreign policy. His most recent books are *Two Chinese States, The Chinese Economy Past & Present,* and *A U.S. Foreign Policy for Asia in the 1980s and Beyond.*

Jean C. Oi is associate professor of government at Harvard University and research associate at Fairbank Center for East Asian Research. She received her Ph.D. in political science from the University of Michigan. Her research area is peasant politics in the Maoist and post-Maoist period and Chinese world politics. Her forthcoming book is *State and Peasant in Contemporary China: The Political Economics of Village Government.*

Dorothy J. Solinger is associate professor at the School of Social Sciences at the University of California, Irvine. She received her Ph.D. at Stanford and will be a visiting professor there in the department of political science 1989–1990. She is the author of *Regional Government and Political Integration in Southwest China, 1949–1954: A Case Study*, *Chinese Business Under Socialism: The Politics of Domestic Commerce, 1949–1980*, and editor of *Three Visions of Chinese Socialism*. She was chair of the China and Inner Asia Council, of the Association for Asian Studies from 1987 to 1989.

Ezra Vogel received his Ph.D. from Harvard, where he is Clarence Dillon Professor of International Affairs and director of the Program on U.S.-Japan Relations, at Harvard Center for International Affairs. Former director of East Asian Research Center (Harvard) and chairman of the Council on East Asian Studies (Harvard), he has done extensive research in Japan, Hong Kong, and The People's Republic of China. He has written *Japan as Number One* and other works.

David Zweig received his Ph.D. in political science from the University of Michigan. He is an assistant professor of international politics at the Fletcher School of Law and Diplomacy at Tufts University. He has written *Agrarian Radicalism and the Chinese Countryside, 1968–1981*, and *China's Agricultural Reform: Background and Prospects*, co-authored with Steven Butler. He was a visiting research scholar at Nanjing University and former executive committee member of the Joint Centre on Modern East Asia, University of Toronto and York University.

INDEX

allocative efficiency, 171-172, 175, 178, 199 n.8, 217, 221
Anhui, 33, 84n.13, 157, 175, table 176, table 177, table 179, 267, 293, 302, 302.

bankruptcy, 27, 29, 35, 149-151, 165n.31, 237, 247-250, 273, 278 (nn. 43,47)
banks, 24-25, 29, 61, 76, 83n.3, 107, 159, 188-190, 192-193, 20in.33, 218, 222-223, table 224, 227, 238, 243, 251, 253-254, 156-259, 263, 274, 275n.10, 276n.21, 280 (nn.76, 78), 291, 295, 298, 305.
Beijing (Peking), 134n.70, 151-152, 157, 159-160, table 176, 235, 238, 242-244, 258, 275n.12, 293, 300.
bonus payments, 24, 37n.8, 72-73, 156, 250, 255.

capital construction, 155, 216, 218-219, 221-222, table 224, 226, 230n.14, 287, 295, 298.
Central Committee, table 18, 44, 89, 124n.1, 173, 299
central planning, 11, 15, 106, 123, 164n.16, 217, 220, 238, 297.
Changchun, 20.
chemical fertilizer, 52, 173, 193, 198n.4.
Chongqing (Chungking), 83n.3, 153, 156-157, 238, 248, 258, 270, 275n.3, 286.
collectivization, 14, 37n.7, 41-43, 61=62.
contracts, 20, 22, 25, 51-52, 54-60, 70-74, 76, 82, 85n.14, 105, 141, 150, 153, 160-161, 164n.16, 172, 178, 183, 193, 195, 249-252, 261, 267, 279n.57, 299, 302.
Cultural Revolution, 11-12, 46, 124n.1.

decentralization, 69, 90, 112, 117, 226-227, 234, 237, 239-241, 244-247, 268, 273-274, 277n.8, 285-287, 295-298.
Deng Xiaoping, 90, 101, 119, 135n.74, 167n.60, 187.

enterprise autonomy, 26, 68, 247-248, 252, 274.
entrepreneurship, 13, 16, 23, 31, 36n.2, 41-42, 57, 62-63, 67, 70, 72, 76, 87n.42, 182, 186.

Fengjia Village, 42-63.
foreign trade, 100-102, 120, 129n.33, 238, 288, 296-298.
Fujian, 41, table 176, table 179, 181, 294, 297.

Gansu, table 176, 190.
government
    bureaucracy, 4-5, 10, 12, 13-16, 20, 23-24, 41, 83n.1, 91, 104, 123, 137, 142, 145, 147-149, 158-161, 220, 233-234, 236, 238, 254, 259, 262-264, 268, 270, 273
    central, 15, 21, 23, 215, 219-220, 222, 228, 241
    local, 15-16, 23-24, 26-30, 33, 35, 72, 75, 77, 79, 81, 147, 188, 219-220, 222, 225-226, 228, 263, 297, 200-301
    village, 69-77, 79, 71.
Guangdong, 10, 11-12, table 176, 186, 200n.26, 297.
Guangzhou (Canton), 11, 153, 157, 238, 275n.3, 276n.16, 284n.142, 297, 305.
Guangxi, table 176, 178.
quanxi, 41, 77, 242, 266-267, 273.
guidance planning, 23, 29, 31, 34, 50, 74-75, 90, 109, 123, 155, 157, 164n.16, 234-237, 240, 245, 251-252, 262, 264, 268-269, 273, 277n.28, 279n.57, 303-304.

Guizhou, *table* 176, 293, 297.

Hebei Province, 20, *table* 176, 287.
Heilongjiang, 175, 193, 288, 292.
Henan, *table* 176, 189-190, 284n.142, 287, 293.
Hong Kong (Xianggang), 11, 147, 157.
housing, 45-46, 113-114, 116, 134n.70, 219, 223, 230n.22, 246
Hu Yaobang, 298.
Hubei Province, 153, 157, 159, *table* 176, *table* 179, 240-242, 245, 247, 258-259, 261, 264, 267, 270, 272, 274, 276n.16, 284n.142.
Hunan, *table* 176, 191, 260, 267, 284n.142.

income
    distribution, 117-119, 123, 134 (nn.72, 73), 135n.74, 144, 146, 159
    equality, 118, 121, 134n.73, 159
    per capital, 9, 44-45, *table* 49, 49-50, 64n.5, 99, 113, 115, *table* 118, 119, 133n.69, 192, 218, 197
    redistribution, 67-68, 78, 81.
inflation, 33, 115, 159, 268.
interest rate, 79, 173, 182, 188-190, 194-197, 202n.56, 256-258.
investment, 2, 4, 6-7, 37n.6, 41, 61-62, 76, 81, 85n.21, 93, 99-100, 102, 105, 109, 111-113, 120-121, 125n.6, 126n.17, 128n.31, 130 (nn.42, 43), 153, 160, 172-173, 193, 205-228, 229n.14, 230 (nn.31, 34), 231n.34, 234, 237, 243, 253-256, 260, 262, 268, 274, 280n.76, 290, 295, 300-302.

Jiangpu County, 17, *table* 19, 20, 25-26, 28, 32-33.
Jiangsu Province, 20, 22, 34, 37n.6, 59 *table* 176, 184, 226, 267, 284n.142, 302.
Jiangxi, 10, *table* 176, 192-193.
Jilin, 44, *table* 176.

labor force participation, 6, 102, *table* 103, 113, 128n.25.
Liaoning, *table* 176, 185, 191, 20 in.32.

market
    bonds, 112, 227, 237, 253-258, 273-274, 298
    capital, 2-3, 4, 12, 14, 37 n.6, 42, 56, 102, 107, 111, 130n.44, 144-148, 153, 155, 237, 243, 249, 253-254, 256-259, 261, 273-274, 275n.10, 28n.10, 28n. 91, 283n.135, 291
    credit, 69, 78, 172-173, 181-183, 187-197, 203n.56
    input, 69, 78, 105, 107-109, 111, 113, 120, 130n.44, 121n.44, 141, 145, 147, 149, 158, 169-203
    labor, 6, 14, 25-26, 49, 102 *table* 103, 109, *table* 110, 111, 113, 146, 172-173, 175, 178, 180, 183-188, 196-197, 199n.14, 201n.32, 213, 217, 293
    land, 14, 29, 172-183, 188, 196-197, 227
    materials, 76, 221, 242-243, 245-246, 249, 252, 254, 261-262, 264-265, 268-271, 273-274, 282n.112, 283n.126, 292, 294, 300, 302, 304.
material incentives, 62, 114, 117, 121.
Mongolia, Inner, *table*176, 291, 302-303.

Nanjing (Nanking), 17, *table* 18, 19-21, 23, 25-26, 29-32, 36n. 4, 258.
national income, 10, 92, 94-99, 101, 107, 126n.7, 128n.25, 129n.33, 169, 198 (nn. 1, 6), 199n.7, 215, 218, 291.
Ningxia, *table* 176, *table* 179, 291-292.

October 1984 Decision, 299
output
    agricultural, 7-8, 21, *table* 32, 43-44, 53-54, 68-69, 85n.11, 86n.41, 95, *table* 97, 104, 108-109, 170, 173, 186, 188, 192, 194, 198n.1, 209, 215, 219, 224, 226-227, 288

chemical fertilizer, 193, 198n.4
cotton, 44, 50, 242, 279n.63
grain, 21, *table* 32, 34-35, 44, 49-50, 79, 193, 243, 288, 292, 294
industrial, 21, 44,49, 8-69, 96 *table* 97, 107, 109, 227, 239, 269, 383n.126, 291, 293
steel, 239, 242, 246, 260, 264, 268-270, 274, 279n.63, 283n.124, 288, 302.

Peng Zhen, 150.
people's commune, 29, 188.
prices
agricultural, 24, *table* 48
free market, 16, 60, 127n.21, 181, 214, 268
grain, *table* 48, 229n.4
state/fixed, 75, 105, 130n.42, 246, 267-268, 270
wholesale, 105.
private ownership, 11, 17-23, 25, 27-28, 35, 36 (nn.3, 4), 51, 63, 68, 90, *table* 103, 147, 152, 157, 196.
private plots, 48, *table* 49, 169, 177-178, 180
production team, 72-73, 169-170, 174-175, 177-180,. 186-187, 190-191.
production brigade, 30, 36n.3, 37n.7, 187.
productivity, 6, 51-52, 93, 106-112, 114, 117, 120, 130n.55, 131 (nn. 44, 49), 132 n.56, 141, 143-144, 146, 149, 153, 158, 170, 173, 185, 199 n.6, 211.

Qingdao, 37n.6, 275n.3.
quotas
compulsory, 22, 29, 298
grain, 29-30
production, 16, 29, 50, 268, 290.

rent, 70-73, 76, 144, 152, 174, 180-182, 184, 186, 197, 200 (nn. 23, 26), 249.
responsibility system
household, 14, 26, 59, 69, 103, 108, 170-174, 177, 184, 188, 191-192, 195-196, 199n.8

management, 26, 155, 157, 237, 247\253, 273, 279n.59
production, 170, 214, 217.

savings
forced, 99, 121, 205-216
household, 62, 230n.14,
rate of, 61, 93, 99, 113, 117, 126n.17, 212, table 216, 217-218.
Shaanxi, *table* 176.
Shandong, 42, 44, 50, 61-62, 64n.5, 84n.13, *table* 176, 287, 293.
Shanghai, 129n.33, 133n.67, 153, 174, *table* 176, 186-187, 238, 256, 258, 260, 270, 276n.16, 286, 288, 293, 297, 300-303, 305.
Shanxi, *table* 176, 188-189, 293, 300, 302.
Shenyang, 68, 72, 80, 83n.3, 86n.41, 150-151, 153, 156-157, 160, 248, 256, 270, 275n.3, 276n.16, 193.
Shulu, 20.
Sichuan, 77, 79, 156, *table* 176, 178, 293-294.
specialized households, 20, 58, 80, 82, 103, 129n.38, 179-181, 183.
State Council, 84n.11, 151, 154, 173, 220, 227, 255, 268, 293, 304.
stock ownership, 150, 152-154, 166n.41, 237, 253-258, 273-274, 180n.79.
surplus labor, 69, 84n.11, 109, 173, 196, 205.
Suzhou City, 34.

Tangquan, 17, 20, 25, 27, 30.
tax
agricultural, 52, 68, 80
value-added, 263, 270.
Third Plenum, *table* 18, 44, 89-90, 93, 95, 124n.1.
Tianjin (Tientsin), 73, 75, 77, 79, 83n.3, 85n.22, *table* 176, *table* 179, 180, 270, 276n.16, 193, 297, 300, 302-305.
township and village enterprise, 36n.2, 51, 53-59, 68-71, 79, 84n.9, 84n.18, 86n.41, 108-109, 129n.38, 219, 226.

trade
  private, *table* 32.
  retail, 102, 104, 266
  wholesale, 24, 29, 33-34, 55, 239, 266, 271-272.

unified distribution, 302.

Wuhan, 153, 158-159, 233-248, 250, 253-256, 258-263, 265-267, 269-273, 275*n*.12, 276 (*nn*. 16, 21), 277*n*.28, 278*n*.47, 279 (*nn*.54, 59, 63), 280*n*.76, 286, 293.
Wujiang County, 34.

Yao Yilin, 297, 303.
Yunnan, *table* 176, 291, 293.

Zhao Ziyang, 90, 119, 239, 295-296, 304.
Zhenjiang, 175, *table* 176, 181, 186, 226, 284*n*.142, 302.
Zhou Enlai, 101.
Zijingshan, 25-26.
Zouping, 42-44, 46, 49, 51-52, 55, 62.